EINSTEIN

THE MAN AND HIS ACHIEVEMENT

EINSTEIN

THE MAN AND HIS ACHIEVEMENT

Edited by

G. J. WHITROW

DOVER PUBLICATIONS, INC.
NEW YORK

This Dover edition, first published in 1973, is an unabridged and unaltered republication of the work originally published by the British Broadcasting Corporation in 1967. It has been reprinted by special arrangement with the British Broadcasting Corporation, 35 Marylebone High Street, London W1.

International Standard Book Number: 0-486-22934-3
Library of Congress Catalog Card Number: 72-98113

Manufactured in the United States of America
Dover Publications, Inc.
180 Varick Street
New York, N. Y. 10014

CONTENTS

FOREWORD

I

The three broadcast programmes which make up the contents of this book have an origin in Israel. In the summer of 1964, during the course of some research which he was doing there, Mr David Tutaev met a certain number of scientists, and in conversation with them the suggestion arose that the BBC ought to do a programme about Einstein and his work now, while there were people available who had known him personally and would no doubt be glad to record their impressions both of the man and his achievement.

Mr Tutaev brought the suggestion to me shortly afterwards when he was back in England. I in turn put it before Mr P. H. Newby, Controller, Third Programme, and his only criticism of the suggestion was that it ought to have been put up and acted on long before. I was delighted at this ready acceptance, but also distinctly alarmed at being put in charge of producing a programme on a subject of which my knowledge was, and remains, extremely meagre. I need not have worried, however, and I quickly ceased to worry after Dr Gerald Whitrow of the Imperial College of Science and Technology took on the general editorship.

The events of Einstein's long life from 1879 to 1955 fall naturally into three sections, first from his childhood to the enunciation of the Special Theory of Relativity in 1905, then the middle years of his life when he became one of the most famous men in the world, and finally the last years, which are here reckoned from 1933, in which his main purpose was to search for the Unified Field Theory which eluded him. Gerald Whitrow was the principal contributor to the first section, Dr Denis Sciama of Cambridge University to the second one, and Professor William Bonnor of Queen Elizabeth

College, London University to the third. Thus the enormous subject was divided 'in tres partes'.

The programmes were first of all intended for 1965, the fiftieth anniversary of the enunciation of the General Theory of Relativity, but it took longer than we expected to get all the contributors we wanted, and so the venture was postponed till 1966. I think that was just as well. Centenaries or demi-centenaries suggest tributes and there is no call to pay a conventional tribute to Einstein. The purpose behind these programmes was not solemn worship, but to take the opportunity which was offered by the lapse of ten years, and the survival of many of Einstein's associates, to look at him in perspective, even if necessary (to borrow the fine phrase which Logan Pearsall Smith used in his book on Shakespeare) 'to tap the pedestal of this imposing statue'.

In fact, in all the contributions to these programmes, the pedestal stands firm. The reason is not that any of the contributors felt inhibitions. Our age is prone to criticism, even to excess, and it is true to say that every contributor to these programmes is of critical mind. Yet there is not one of their contributions which shows any tendency to take away from the enormous reputation which was accorded to Einstein when his name first became famous in the world at large in 1919.

All broadcasting on matters of science has one aim, that of making the appreciative public wider. Here our three programmes meet their gravest challenge. Can it be said that these three hour-long programmes, even when reprinted as here *in extenso*, and including a few passages sacrificed in the broadcast version to the exigencies of time – can such an enterprise really have the desired effect of increasing the numbers of those who understand the importance of Einstein's achievement as a scientist?

It is true that to appreciate fully what Einstein did it is necessary to have a mathematical equipment which only a few can make claim to; but Einstein's achievement would have only a limited significance if it was confined to specialist interest and study. Let me give this quotation from a note by Gerald Whitrow:

'Einstein's influence was not confined to the technicalities of modern physics. For, just as it is inconceivable that there will be any general reversion to pre-Copernican, pre-Newtonian or pre-Darwinian assumptions concerning the general nature of the universe and man's place in it, there will likewise be no return to the world-view of Einstein's predecessors.'

But there remains that disturbing question as to whether any appreciation at all of Einstein is possible without intense previous training. The example of the past may give us hope however.

When Newton formulated the first successful theory of gravitation, his notion of a force of gravity, so familiar to everyone with any education today, struck the majority, including many eminent scientists, as completely incomprehensible. It is true that most people today (such as myself) would be incapable of giving the merest outline of Newton's theory, but they would not regard it as incomprehensible. They would need notice of the question, that is all. It was much the same at the end of Newton's own time. From this it is clear that the interval between the enunciation and the general adoption of his ideas as part of the accepted mental furniture without which a man can have no claim to be educated was very much less in Newton's case than in Einstein's. In the immense and diverse scientific activity of the present time, the greatest age of science the world has known, digestion must take longer because there is so much more to be digested. This means that today we need aids to digestion to a degree which was not necessary in a relatively simpler time. The aids need to be more elaborate than was the case formerly. It is impossible to give a popular version of Einstein's theories, such as makes them immediately accessible today to everyone, without distorting them, but it is possible to convey their content in language which is not specialist and makes them accessible in outline to those who are prepared to give close attention. That was the humble and, as I dare to say, the vital purpose of the radio programmes which are here reprinted. To succeed in such a purpose should not be impossible.

The past history of science can give us encouragement, but all history warns against the dangers of complacency. It is by no means inevitable that all attempts to popularize science or philosophy, essential to civilization as such attempts are, will be successful. We have evidence from the past of how extraordinarily short cultural memory can be. We should remember that three hundred years after Eratosthenes of Alexandria had made a remarkably close measurement of the circumference of the earth, the polished Tacitus, living in a refined Hellenized society in Rome at the height of the Empire, evidently believed that the earth was flat!

II

Radio-programmes such as these which rely on a great deal of first hand memory are inevitably biographical. Here we may help to fill a gap, as no definitive biography of Einstein has yet appeared. One reason may be that he is not an easy subject for biography. It emerges from several contributions that he was a man of lovable personality and honourable character, but at the same time it is clear, most notably, I think, in what his son H. A. Einstein has to tell, that he depended far less than most men on personal relations. He was not a man of narrow interests but, as is conveyed forcibly by Sir Roy Harrod in his memories of Einstein as a 'Student of Christ Church', his mental energy was fully committed to the pursuit of science, outside which, Sir Roy tells us, 'I found that his views were liberal and enlightened, but not particularly deep. He evidently accepted what was in current circulation without bringing that great brain of his much to bear on such topics.'

With his broad sympathies he was evidently a man of robust humour, as the contributions of Miss Margaret Deneke and Professor Bernard Cohen make clear. A nice example of his humour was the German doggerel verse he left behind as his letter of thanks to the late Robin Dundas, whose rooms he occupied in Oxford in 1931. (There was not time in the programme for this comic poem which is preserved in the Bodleian Library.) The most perfect example of his tact and sensitiveness is the story given in the second section by Mr L. L. Whyte about Einstein's conduct when he discovered that his collaborator Cornelius Lanczos had worked out an equation given to him in error. Yet in spite of so many touches which show his friendliness there is every sign that he was extraordinarily self-sufficient. Only a man as self-sufficient as he, could have worked out his first epoch-making discoveries in obscurity. But despite his friendships he was essentially a lonely figure. It was perhaps a penalty he had to pay for an endowment of genius of a magnitude which appears but rarely in the whole of recorded history. It may be that a further and lesser penalty is that there will be no biography of Einstein commensurate with its subject.

In his last contribution in the third section, Gerald Whitrow says that 'Einstein was no doubt a somewhat more complex personality than generally imagined'. Even so, the lasting impression of Einstein is likely to be that of a man endowed with that kind of simplicity which belongs to greatness alone.

Nevertheless there are some myths about Einstein that need to be cleared away, and these programmes should help in that direction. The main and most pernicious one concerns the last phase of his life and his failure to enunciate a Unified Field Theory. Compared to the astounding successes of his early and middle years any failure of Einstein tends to look dark and terrible indeed, and as a result there has been an unfortunate tendency to consider him in his old age as a man of enfeebled energy and even decaying mind. In fact his mind was resilient and inventive to the last. Professor Bergmann, whom I had the privilege of interviewing in New York in 1965, is convinced that ultimately science will have to evolve a Unified Field Theory, and that if one is going to criticize Einstein it should not be for bending his energies in that direction, but for not recognizing that such a theory as he hoped to enunciate was not possible with the information then (and now) at the disposal of scientists. If Einstein incurs blame it is not for enfeeblement but for boldness.

In this connection I may close by recalling an interview which was given to me at Berkeley, San Francisco, by Professor Taub. Unfortunately there was not time to include the recording in the third part. Professor Taub told me that he much regretted that Einstein dedicated himself to what seemed to him the hopeless quest of the Unified Field Theory, and since he was on close terms with him he asked Einstein one day why he was using his gifts in that way. Einstein replied that he agreed that the chance of success was very small but that the attempt must be made. He himself had established his name; his position was assured, and so he could afford to take the risk of failure. A young man with his way to make in the world could not afford to take a risk by which he might lose a great career, and so Einstein felt that in this matter he had a duty. Such was his answer.

Can one imagine a more perfect example of the selfless devotion of the inquiring mind informed by the sympathy of a character instinct with generosity?

March 1967   CHRISTOPHER SYKES

BIOGRAPHICAL
NOTES ON CONTRIBUTORS

DR G. J. WHITROW is Reader in Applied Mathematics in the University of London at the Imperial College of Science and Technology. Formerly Lecturer of Christ Church, Oxford. Vice-President of the Royal Astronomical Society, 1965–7. Author of *The Natural Philosophy of Time*, *The Structure and Evolution of the Universe*, etc.

PROFESSOR ANDRÉ MERCIER is Head of the Department of Theoretical Physics in the University of Berne, Switzerland and Secretary-General of the International Committee on General Relativity and Gravitation.

BERNARD MAYES is the BBC Correspondent in California and the West Coast of the USA.

PROFESSOR H. A. EINSTEIN, elder son of Albert Einstein, is Professor of Hydraulics in the University of California at Berkeley.

BERTRAND RUSSELL (Earl Russell), O.M., F.R.S., won the Nobel Prize for Literature in 1950. His many books include *The ABC of Relativity*.

SIR KARL POPPER, F.B.A., is Professor of Logic and Scientific Method in the University of London at the London School of Economics. Author of *The Logic of Scientific Discovery*, *Conjectures and Refutations*, etc.

DR R. H. FURTH was formerly Reader in Physics in the University of London at Birkbeck College and at one time Professor of Physics in the Charles University, Prague.

PROFESSOR OTTO FRISCH, F.R.S., is Jacksonian Professor of Natural Philosophy, Cavendish Laboratory, University of Cambridge. Famous for his pioneer work on nuclear fission.

DR D. W. SCIAMA is Fellow of Peterhouse and Lecturer in Applied Mathematics in the University of Cambridge. Author of *The Unity of the Universe*.

PROFESSOR H. LEVY, F.R.S.E., is Professor Emeritus of Mathematics, Imperial College of Science and Technology, University of London. Author of *The Universe of Science*, etc.

PROFESSOR HERBERT DINGLE is Professor Emeritus of the History and Philosophy of Science, University College, University of London. Past President of the Royal Astronomical Society. Author of *The Scientific Adventure*, *Relativity for All*, etc.

PROFESSOR W. H. MCCREA, F.R.S., is Director of the Astronomy Centre in the School of Mathematical and Physical Sciences, University of Sussex. Past President of the Royal Astronomical Society. Author of *Relativity Physics*, etc.

PROFESSOR CORNELIUS LANCZOS is Professor in the School of Mathematical Physics in the Dublin Institute for Advanced Studies. Author of *Albert Einstein and the Cosmic World Order*, etc.

DR E. H. HUTTEN is Reader in Physics in the University of London at the Royal Holloway College. Author of *The Language of Modern Physics*, etc.

L. L. WHYTE is Past President of the British Society for the Philosophy of Science. Author of *Critique of Physics*, etc.

MARGARET DENEKE is Honorary Fellow of Lady Margaret Hall, Oxford. Has lectured widely in Great Britain and the USA on musical subjects.

SIR ROY HARROD, F.B.A., Student of Christ Church, Oxford, is one of the leading economists in the country. Author of *The Prof: a Personal Memoir of Lord Cherwell*, etc.

PROFESSOR W. B. BONNOR is Professor of Mathematics in the University of London at Queen Elizabeth College. Author of *The Mystery of the Expanding Universe*.

PROFESSOR PETER BERGMANN is at the University of Syracuse, New York. Scientific assistant to Einstein at the Institute for Advanced Studies, Princeton, 1937–8.

PROFESSOR BANESH HOFFMANN is at Queen's College, Flushing, in the City University of New York. Collaborated with Einstein and Leopold Infeld, 1937–8. Author of *The Strange Story of the Quantum*, etc.

PROFESSOR J. A. WHEELER is at Princeton University, New Jersey. President of the American Institute of Physics, 1966. Author of *Geometrodynamics*, etc.

PROFESSOR ERNST STRAUS is in the University of California at Los Angeles. Scientific assistant to Einstein at the Institute for Advanced Studies, Princeton, 1944–7.

PROFESSOR NATHAN ROSEN is at the Israel Institute of Technology, Haifa. Scientific assistant to Einstein at the Institute for Advanced Studies, Princeton, 1934–5.

PROFESSOR HERMANN BONDI, F.R.S., is Professor of Mathematics in the University of London at King's College and Chairman of the Space Committee, Ministry of Defence. Author of *Cosmology*, *Relativity and Common Sense*, etc.

PROFESSOR I. BERNARD COHEN is Professor of the History of Science at Harvard University. Author of *Newton and Franklin*, etc.

DR MEYER WEISGAL is President of the Weizmann Institute, Rehovoth, Israel.

PROFESSOR EUGENIUSZ OLSZEWSKI is Professor of the History of Science and Technology at the Warsaw Institute of Technology.

I

EARLY YEARS AND FIRST ACHIEVEMENTS
1879-1905

Whitrow: Albert Einstein, one of the greatest physicists of all time, was born of Jewish parents in the town of Ulm in southern Germany on 14 March 1879. Unfortunately, no trace survives of the house in which he was born. It vanished in the artillery and bombing attack of 1945. In 1922, when Einstein had already attained world fame, an unimportant street in Ulm was named after him, but following Hitler's rise to power in 1933 it was renamed after Fichte, the early nineteenth-century philosopher of German nationalism. Since 1945 it has again been called Einsteinstrasse.

When Albert was a year old the family moved to a suburb of the Bavarian capital Munich, where Albert's father, Hermann Einstein, ran a small electrochemical factory with his brother, who lived with the family. There was one other child, a daughter Maja, one year younger than Albert.

As is so often the case with men of genius, Einstein's father was a somewhat ineffectual person whose business was frequently unsuccessful. His mother, born Pauline Koch, was the stronger character. She had a genuine interest in German classical music, particularly Beethoven's piano sonatas, but the main intellectual stimulus in Einstein's early life came from his uncle, who was a trained engineer.

Like Newton, Einstein showed no early signs of genius. On the contrary, he was slow in learning to talk and as a child he was rather taciturn. He did not play much with other children, but tended to be pensive and a day-dreamer. To most people he must have seemed a rather dull little boy. Unlike most children, he particularly disliked playing at soldiers and all his life he deplored the coercion of one group of people by another. In his early years he was attracted by traditional religion and could not accept his father's scepticism. This religious feeling later found expression in a profound belief in the essential harmony of the laws of nature.

Whitehead has said that philosophy begins with wonder, and Einstein's first recorded experience of his sense of wonder at the marvels of nature seems to have occurred when he was about four or five. His father showed him a magnetic compass. The sight of the needle pointing always in the same direction, whenever the compass case was turned about, made a deep and lasting impression on him. He realized that there must be something that caused the needle to turn in a particular direction, but as there was no visible agent he concluded later that it must be some mysterious property of apparently empty space.

When he was ten, Einstein was sent to the Leopold Gymnasium in Munich, where most of the time was devoted to studying Latin and Greek grammar. Einstein found the mechanical learning of language irksome, and the only teacher who aroused his interest did so by introducing him to the writings of Goethe and Schiller. An amusing story is told about Einstein and this teacher. Years later, when he was already a professor in Zürich, Einstein paid a visit to Munich and decided to call on his former teacher. Einstein was rather shabbily dressed, and the teacher, who had no clear recollection of him, could only imagine that by claiming to be one of his former pupils the young man thought he could borrow money from him. The result was that Einstein soon left, thoroughly embarrassed!

Einstein's interest in mathematics was first aroused at home and not at school. His uncle introduced him to algebra, and he found much pleasure in solving simple problems by ways of his own instead of slavishly following a standard method. When he was about thirteen or fourteen his attention was drawn to a series of popular books on natural science which he absorbed with youthful enthusiasm. At about this time his attitude towards religion became more clearly defined. Although he rejected religious ritual and decided not to join any religious group, he became convinced of the great ethical value of the Biblical tradition.

While the young Einstein was becoming more and more absorbed in his reading of science and in the study of mathematics, his father's business difficulties were increasing to such an extent that in 1894 he decided to leave Munich and seek his fortune in Italy. Albert was left behind in a pension to complete his studies in the gymnasium, but after six months he could stick the life no longer and he joined his parents in Milan. It is said that his departure, for which he contrived to obtain a medical certificate, was facilitated by his teacher telling him that he was a disruptive influence in the class!

He was delighted to escape from the rigours and discipline of life in Germany and move among people who behaved more spontaneously. But the problem now arose of his higher education. His interest lay in theoretical physics, but without a diploma from the gymnasium it was difficult to enter a university. He therefore sought admission to the famous Polytechnic Institute in Zürich, but he failed the entrance examination, despite his good showing at mathematics. Instead, he was advised to attend an excellent school in Aarau and obtain a diploma. Fortunately, the atmosphere in this school appealed to him much more than that of the gymnasium in Munich, and in 1896, at the age of seventeen, he was at last admitted to Zürich Polytechnic (Federal Institute of Technology) as a student of mathematics and physics.

A classmate at the cantonal school in Aarau which Einstein attended between his two sittings of the Polytechnic entrance examination later said of him: 'Unhampered by convention, his attitude towards the world was that of a laughing philosopher and his witty mockery pitilessly lashed any conceit or pose. In

3

conversation he always had something to give. His well-schooled taste acquired from travelling – his parents lived at that time in Milan – gave him a maturity of judgement. He made no bones about voicing his personal opinions whether they offended or not. This courageous love of truth gave his whole personality a certain *cachet* which, in the long run, was bound to impress even his opponents.' Einstein loathed any display of sentimentality, and the same companion concluded that he was one of those split personalities who know how to protect with mocking wit an intense emotional life.

The four fellow students who enrolled for the same course as Einstein included Marcel Grossmann, who played a vital role in later years in drawing his attention to the kind of mathematics required for the extension of the theory of relativity to gravitation, and Mileva Marič, a Serbian whom Einstein married in 1902.

At Zürich, Einstein soon realized that he was by nature a theoretical physicist rather than a mathematician. Like most men of genius, he was largely self-taught. He wasted little time on compulsory lectures, which he often cut, but studied privately, with his girl friend Mileva, the great classics of theoretical physics, the works of Helmholtz, Hertz, Kirchhoff, Boltzmann and, above all, Clerk Maxwell. It is perhaps not surprising that the rather conventional and uninspiring professor of theoretical physics, Heinrich Weber, disliked Einstein and prevented him from becoming an assistant after he received his diploma. The fact that for the examinations he was compelled to burden his mind with a mass of detailed factual knowledge had such a bad effect on Einstein that after his finals in 1900 he lost interest in scientific problems for a whole year. As he wrote in his autobiography many years later, 'It is nothing short of a miracle that modern methods of teaching have not yet entirely strangled the sacred spirit of curiosity and inquiry; for this delicate plant needs freedom no less than stimulation.'

Einstein's lack of reverence for some of his teachers was not due to intellectual arrogance and conceit. He revered the achievements of his great predecessors and he was influenced by the lectures of the famous mathematician Hermann Minkowski, who held a chair in the Zürich Polytechnic from 1896 to 1902. In the summer term of 1900, Einstein assiduously attended his lectures on analytical mechanics. Outside the ranks of pure mathematicians, Minkowski's fame today rests on his invention in 1907 of the concept of space-time, which had such a decisive influence on Einstein's development of general relativity and is now regarded as one of the most important contributions ever made by a mathematician to natural philosophy.

Although Einstein's teachers failed to recognize his genius – even Minkowski admitted that the publication in 1905 of the special theory of relativity surprised him, since, to quote his own words, it seemed to him that 'in his student days Einstein had been a lazy dog' – his fellow student Marcel Grossmann was far more perceptive. A few days after their first meeting he said to his parents, 'This Einstein will one day be a very great man.'

Apart from theoretical physics, Einstein had a passion for music. The daughter of his Zürich landlady tells a charming story. One summer day he heard someone playing one of Mozart's piano sonatas in a neighbouring house and asked who it was. 'I told him that it must be a piano teacher who lived in the attic. He hurriedly put his violin under his arm and rushed out without collar or tie. "You can't go like that, Herr Einstein," I cried, but either he did not hear or pretended not to hear me. A moment later the garden gate banged and it was not long before we heard a violin accompanying the Mozart sonata. On his return Einstein said with great enthusiasm, "That's a really charming little old lady. I shall often go and play with her." We were to meet a few hours later. It was old Fräulein Wegelin who soon appeared in a black silk dress and asked shyly the name of this extraordinary young man. We pacified her by saying that he was merely a harmless student. She told us what a shock it had given her when the unknown musician rushed into her room and merely said, "Go on playing." '

Einstein hated most things that other men hold dear. 'Comfort and happiness,' he declared in later life, 'have never appeared to me as a goal. I call these ethical bases the ideals of the swineherd. . . . The commonplace goals of human endeavour – possessions, outward success and luxury have always seemed to me despicable, since early youth.' He was essentially a lone wolf. He never took part in any student gatherings, but although he rejected the churches he had a Spinoza-like belief in a cosmic religious force. He regarded this as an eternal spiritual being that communicates small details of itself to our weak and inadequate minds. As he once declared, 'This deep intuitive conviction of the existence of a higher power of thought which manifests itself in the inscrutable universe represents the content of my definition of God.' In other words, he had no more use for the shallow materialism that is the most widely accepted philosophy of scientists and others today than for the authoritarian views of the churches that were once so powerful. Of all religious bodies, the one that he felt most sympathy for was the pacificist Society of Friends, the Quakers. And in this connection it is perhaps not altogether irrelevant to mention that the first eminent scientist to expound Einstein's theory of relativity in this country was, in fact, a Quaker, the late Sir Arthur Eddington.

The outlook for Einstein after graduating was bleak. An aunt in Genoa who had made him a monthly allowance ceased to pay it. Soon afterwards his father died and his mother had to work as a housekeeper. As one of his biographers has said, 'That he managed to maintain himself financially in Zürich until the autumn of 1901 can be attributed to Einstein's stoical indifference to material things.' He did some calculations for the director of the Federal Observatory and then taught for a few months in a technical school in Winterthur while the resident teacher of mathematics was doing his military service. Meanwhile he became a Swiss citizen of Zürich, and as he never repudiated it he retained this citizenship until his death. He was never called for military service, however, because he failed his medical on account of flat feet and varicose veins, a verdict that irritated him!

At last, in 1902, his friend Marcel Grossmann introduced him to the director of the Swiss Patent Office in Berne, who decided to appoint him to his staff, in June of that year, at a salary that enabled him to live quite comfortably. Soon afterwards he married Mileva. Although the marriage seems to have been reasonably happy at first, it eventually broke up. They had two sons.

Einstein's move to Berne was a turning-point in his life. Although he had had no previous experience with technical inventions, he found the work in the Patent Office interesting. It was his duty to put applications into a clear form and to determine the basic idea from the often vaguely worded descriptions of the inventors. It may well have been this training that developed his remarkable faculty for seeing to the heart of a problem and quickly realizing the consequences of any hypothesis. Moreover, the work left him with ample time to pursue his own ideas. Indeed, it would seem in many ways to have been the ideal post for him at this stage of his career.

Einstein's first scientific paper appeared in 1901. It was on capillary attraction. Among other early papers were two, published in 1902 and 1903, on the statistical foundations of thermodynamics, a theory that had made a profound impression on him. His work amounted to a new derivation of the main features of statistical mechanics by a similar method to that developed a year or two earlier by the American mathematical physicist Willard Gibbs, although Einstein was unaware of this. Einstein's approach was less abstract than that of Gibbs and he went on to consider a practical application of the greatest significance.

At this time the reality of molecules and of the kinetic theory of matter, according to which the temperature of a body is due to the thermal agitation of its constituent molecules, were still under dispute. Einstein discovered that this thermal agitation can produce a visible and measurable effect on particles suspended in a solution. Such an effect had, in fact, been discovered by the Scottish botanist Robert Brown as long ago as 1827, when he observed that pollen grains suspended in water became dispersed in a great number of small particles which were in constant motion, moving in irregular zigzags even in the absence of currents and other external disturbances.

The fundamental paper in which Einstein showed that the Brownian motion could be used as direct evidence for the existence of molecules and the correctness of the kinetic theory of heat was published in 1905. He argued that, although the velocity of a suspended particle due to the impacts of the molecules of the liquid on it is unobservable, the effect of a succession of irregular displacements can be detected with the aid of a microscope. The actual observations were made later by the French physicist Jean Perrin and completely confirmed Einstein's theory. This pioneer analysis of a phenomenon of thermal fluctuations has been the inspiration for many later investigations of similar effects. Moreover, the phenomenon of Brownian motion has come to be regarded as one of the best 'direct' proofs of the existence of molecules. Einstein's researches on this subject also played their part in helping to convince physicists of the importance of probability in relation to natural laws. Nevertheless, Einstein himself seems always to have believed that the ultimate laws are essentially causal and deterministic and that it is only our inability to deal with large numbers of particles in any other way that compels us to use statistical methods.

The year 1905 was Einstein's *annus mirabilis*. Because of his contributions, volume 17 of the *Annalen der Physik* of that year is now regarded as one of the most remarkable volumes of scientific literature ever published. A copy was recently offered for sale by a London dealer at no less than £550! It contains three papers by Einstein. Each is on a different subject and is a masterpiece. The paper on the Brownian motion was the second of these. It was preceded by Einstein's first contribution to quantum physics. Like the third paper, which was on relativity, it was concerned with the behaviour of light.

One of the problems that puzzled physicists at the end of the nineteenth century concerned the light radiated from a hot body. As the temperature rises, the body changes in colour from a dull glaring red to a brighter orange and then to a blinding bluish white. Attempts to explain this change in the quality of light with temperature had failed to show why at a given temperature there is no perceptible radiation above a certain frequency of vibration corresponding to light of a definite colour. It appeared that it must be difficult to emit light of very high frequency, but no one knew why. To resolve the problem Max Planck assumed, in 1900, that the energy of radiation emitted with a given frequency ν by an oscillating atom cannot be less than a definite minimum amount $h\nu$ and must increase by units of this amount, where h is a new universal constant of nature, now known as Planck's constant. Thus light of a given frequency, or colour, can only be emitted and absorbed in discrete packets of energy, or quanta, and not in purely arbitrary amounts. With this assumption, Planck derived results in the theory of radiation that agreed with observation, and even succeeded in calculating the value of his new constant h.

Planck's hypothesis was slow to make its effect felt on physics, because it seemed difficult to reconcile with the generally accepted wave-like character of radiation. Planck still believed that light and all other forms of electromagnetic radiation were continuously divisible and that the quantum effects concerned only their interaction with matter. This contrast between the continuous nature of radiation and the discontinuous character of its emission and absorption by atoms puzzled Einstein. He came to the conclusion that radiation itself must have a corpuscular structure and that it is actually *composed* of Planck's quanta. He showed that this hypothesis provided a simple explanation of a curious physical phenomenon, known as the photoelectric effect, that had been a great puzzle to physicists for some years. It had been found that when light shines on metallic bodies electrons are emitted if, and only if, the frequency of the light exceeds a certain threshold value depending on the particular metal. Moreover, the speed with which the electrons are emitted does not depend on the intensity of the light, but only on its frequency, or colour.

Einstein showed that these observations could be explained if the energy in a beam of light of frequency ν were concentrated in small packets, each of energy $h\nu$ where h is Planck's constant. These packets of radiation are now called photons. For an electron to be ejected a certain amount of energy must be expended, depending on the nature of the metal. Consequently, if a bombarding photon is to liberate an electron from the metal, it must have sufficient energy and hence its frequency must be above a certain threshold value. Moreover, the speed with which the electron is ejected will depend on the excess energy of the incident photon, and this in turn will depend on the amount by which its frequency exceeds the threshold value. Einstein's simple formula for the speed of the emitted electrons was later confirmed experimentally and a value obtained for h in good agreement with Planck's original estimate.

In introducing his quantum hypothesis, Planck seems to have thought that he was making only a minor adjustment to the classical laws of physics. Einstein believed that a much more radical point of view was necessary and that the whole structure of mechanistic physics based on Newton's laws of motion was at stake. This belief found further expression in the third of his great papers of 1905. It originated in another problem concerning the behaviour of light and gave rise to what is now known as the special theory of relativity.

It happens that we have detailed evidence concerning the way in which Einstein gradually came to develop this theory. For, in 1916, the distinguished psychologist Max Wertheimer had several long conversations with Einstein on this question. He afterwards presented a fascinating account in one of the chapters of his book *Productive Thinking*. Already, at the age of sixteen, Einstein had been puzzled by the following paradox. According to generally accepted ideas, a beam of light travels in empty space with a finite velocity of approximately 300,000 kilometres a second. Einstein tried to imagine what he would observe if he were to travel through space with the same velocity as such a beam. According to the usual idea of relative motion, it would seem that the beam of light would then appear as a spatially oscillating electromagnetic field at rest. But such a concept was unknown to physics and at variance with Maxwell's theory. Einstein began to suspect that the laws of physics, including those concerning the propagation of light, must remain the same for all observers however fast they move relative to one another.

When Wertheimer asked Einstein if already at this time he had some idea of the invariance of the velocity of light for all observers in uniform relative motion, Einstein replied, 'No, it was just a curiosity. That the velocity of light could differ depending on the movement of the observer was somehow characterized by doubt. Later developments increased that doubt.'

Nevertheless, as he told Wertheimer, it was only after years of thought that he finally felt compelled to regard the velocity of light as a fundamental invariant independent of the motion of the observer, for this idea conflicted with traditional views concerning the measurement of motion. How, then, is motion to be measured? Einstein realized that it must depend on the measurement of time. 'Do I see clearly', he asked himself, 'the relation, the inner connection between the measurement of time and that of movement?' It occurred to him that time measurement depends on the idea of simultaneity. Suddenly he was struck by the fact that, although this idea was perfectly clear when two events occur in the same place, it was not equally clear for events in different places. This was the crucial stage in his thinking. For he saw that he had discovered a great gap in the classical treatment of time. It took him roughly ten years to arrive at this point, but from the moment when he came to question the traditional idea of time, only five weeks were needed to write his paper, although he was working all day at the Patent Office.

The critical reasoning that led Einstein to abandon the classical concept of world-wide simultaneity was stimulated by his interest in philosophy. Soon after his move to Berne in 1902 he had made the acquaintance of a Rumanian named Maurice Solovine, who was studying both physics and philosophy, and of a Swiss student named Conrad Habicht. In the evenings they often read together and discussed classics of philosophy by Plato, Kant, Mill, Poincaré, and others.

According to Einstein's own account, the philosophers who helped him most to develop his critical powers were David Hume and Ernst Mach. Hume influenced Einstein by his penetrating criticism of traditional common-sense assumptions and dogmas. Mach's influence was more direct and at the same time more complex. For Einstein was not at all in sympathy with Mach's general philosophy of science based on the doctrine that the laws of physics are only summaries of experimental results. Instead, Einstein believed that these laws also involve factors contributed by the human mind. Nevertheless,

Mach influenced Einstein by his criticism of Newton's ideas concerning space and time and also by his critical examination of Newtonian mechanics.

While Einstein was gradually being led to question the classical conception of time, he was also becoming increasingly sceptical of the mechanistic idea that electromagnetic waves in empty space must be regarded as oscillations in a peculiar universal medium called the ether. Indeed, the properties of this medium seemed to defy mechanical explanation.

On the one hand, it seemed clear that the Earth does not drag the ether along with it in its motion around the Sun. This had been shown by Bradley's discovery that the apparent directions of the stars exhibit small annual changes due to the cyclical change in the Earth's direction of motion during the course of the year. For, if the ether in which the light waves from the stars undulate were dragged along by the Earth, no such effect would be observed. Consequently, the Earth must move through the ether.

If, however, the Earth moves through the ether, then it should be possible in principle to determine its velocity relative to the ether by measuring the velocity of light relative to the Earth in different directions. This had been the object of the famous Michelson-Morley experiment of 1887. Various attempts were made to explain its failure, despite the adequacy of the apparatus to detect the effect of a velocity through the ether considerably less than that with which the Earth was known to move around the Sun. Although Einstein's train of thought seems, in fact, to have been but little influenced by this particular experiment, when he eventually heard of it he immediately realized that the trouble was due to introducing the idea of the ether as the medium of the transmission of light. Just as Mach had rejected Newton's concept of absolute space, so Einstein discarded the luminiferous ether and with it the mechanical conception of optical phenomena. In his view, Michelson and Morley obtained a null result because, despite the Earth's motion, their apparatus must have functioned exactly as if it had been at rest throughout the experiment.

A similar state of affairs had long been familiar in mechanics. It was known that mechanical forces produce the same effects on bodies in uniform motion as on bodies at rest. For example, any mechanical experiment performed on board a ship sailing steadily in a straight line yields the same result as a similar experiment carried out on land. Newton's laws of motion had therefore been formulated so as to be the same for all frames of reference in uniform relative motion, including those at relative rest. On this basis, it is impossible, even in principle, to devise a mechanical experiment to measure an absolute velocity.

In 1904 the great French mathematician Henri Poincaré introduced a general physical law to cover both this situation in mechanics and the result of the Michelson-Morley experiment in optics. He called it the 'principle of relativity'. According to this principle, now known as the 'principle of special relativity', the laws of physics are the same for all observers at rest or moving uniformly in straight lines. Nevertheless, although Poincaré's principle implies that, as far as experimental physics is concerned, all uniform motion is relative and no absolute uniform motion is detectable, he did not go on to construct a theory of relativity, as Einstein did shortly afterwards. For, unlike Einstein, Poincaré failed to realize that his principle made the ether concept unnecessary and indeed was not strictly compatible with it. Instead, he continued to believe that the ether must be retained as a mechanical basis for the transmission of light.

The true creator of the special theory of relativity was therefore Einstein in his paper of 1905, in which he accepted the principle of relativity unconditionally as a fundamental general law of physics. As we have already seen, he had been feeling his way towards this principle on what we may call 'philosophical' grounds for many years. He did not attempt to account for it in terms of other physical hypotheses, nor did he appeal to the Michelson-Morley experiment to justify it, but he regarded it as a more suitable starting-point for the general study of physical phenomena than Newton's laws. It was this, more than anything else, that made Einstein's theory so difficult for older physicists to accept. Instead of reducing optics to mechanics, Einstein based both optical and mechanical laws on the same fundamental principle of relativity.

In accordance with this principle, light should have the same properties for all observers and frames of reference in uniform relative motion. In particular, as Einstein was forced to conclude, its velocity in empty space *must* be the same for all of them, despite the fact that this condition clashes with the common-sense assumption of world-wide simultaneity. Einstein therefore abandoned this assumption and explored instead the consequences of using the invariance of the velocity of light as a means of comparing the time readings of clocks in uniform relative motion in different places. He found that, on this basis, different observers would, in general, assign different times to the same event and that a moving clock would appear to run slow compared with an identical clock at rest with respect to the observer. Although these effects are small, except when the relative motion is nearly that of light, they seemed paradoxical and were received with incredulity by many educated people, including most philosophers.

Einstein showed that his principle could be used to derive from the classical laws of physics, assumed to be valid for velocities that are small compared with the velocity of light, laws that are valid for all velocities. In particular, he found that Newton's laws of motion, formerly regarded as the foundation of physics, had to be modified for rapidly moving bodies. Thus the inertial mass of a body, formerly supposed to be independent of motion, was now found to increase indefinitely the nearer its velocity approaches that of light. Consequently, a given force acting on the body will produce smaller and smaller changes in its velocity the faster it moves. As a result no particle of matter can ever attain the speed of light. In this way, Einstein solved his ten-year-old problem concerning the observer who moves with the same velocity as a beam of light. This motion is physically impossible.

The dependence of mass on velocity led Einstein to a remarkable unification of concepts. He came to the conclusion that mass and energy are intimately associated. Corresponding to any increase in the energy-content of a body, there is an equivalent increase in its mass. In a short paper, published later in 1905, Einstein similarly showed that, if a body gives off energy E in the form of radiation, its mass is diminished by an amount M, where $E = Mc^2$.

This unification of mass and energy was one of the most important consequences of Einstein's special theory of relativity. It implied that matter can be regarded as highly concentrated energy. This hypothesis has received remarkable confirmation in nuclear physics and has resolved the problem of the origin of the Sun's radiation. The enormous release of energy in nuclear reactions is due to the conversion of a small quantity of mass into its equivalent large amount of liberated energy.

To Einstein the main value of his discovery concerning mass and energy did not lie in its practical applications. The vital point was that he had come upon it as a consequence of the relativity principle. He believed that this principle was applicable to all natural phenomena and that it could unite the laws of nature. However, before he could successfully apply it to gravitation, he had to generalize the principle to cover accelerated motion, and this problem was to be his main preoccupation for the next ten years.

Although it was by this later work on general relativity

that Einstein was destined to attain world-wide popular fame, his professional standing among physicists can be traced back to the great papers of 1905. For it was in these papers that he first revealed the essential limitations of classical theories and laid the foundations of twentieth-century theoretical physics.

Today there must be very few people who can remember Albert Einstein as a young man in Berne. Professor André Mercier is a distinguished scientist living in that city who is head of the department of theoretical physics in the University of Berne and also Secretary-General of the International Committee on General Relativity and Gravitation. When he was last in London I asked him to tell me something about Einstein's early life in Switzerland.

Mercier: The fact that the young Albert was from his youth brought up in Switzerland certainly had a decisive influence on him. The simple and democratic way of Swiss life was in those days very different from that of neighbouring countries. College years spent in the mainly Protestant environment revealed to Einstein that he did not easily conform to Jewish traditional orthodoxy so that, although in some circles it has become fashionable to insist upon Einstein's Jewishness, he himself never felt, except possibly in his later years, that he primarily belonged to that creed and tradition. He was much too cosmopolitan for that.

Whitrow: Was not Einstein a Swiss citizen?

Mercier: Well, that is an interesting question. When he came to our country as a boy he was by nationality a German and remained so until he became of age; then he expressly applied for Swiss citizenship. Although it is notoriously difficult to acquire Swiss citizenship, it was granted to him. Later when he went to Berlin he was again made a German citizen and many years later, after he had settled at Princeton, New Jersey, American citizenship was conferred upon him by an act of Congress, but these successive nationalities were bestowed upon him almost like honorary degrees. Nevertheless, he retained his Swiss citizenship until the end of his life. In virtue of this, he had a traditionally international neutral status, and he was certainly vividly aware of its significance. In this connection it may be mentioned that the only diploma he had on the walls of his office in Princeton was that of an honorary member of the Berne Society of Sciences. The years Einstein spent in Berne, where he wrote his first revolutionary scientific papers and where he was virtually unknown except to a handful of dear friends, were probably the happiest of his life.

Whitrow: You may have known some of his friends in Berne. Could you describe one or two of them?

Mercier: Well, it is not so easy and I shall not dwell upon his first marriage, which brought him to a tiny flat in the picturesque medieval part of the city. One friend had an important influence on Einstein's early scientific career. This was Michele Besso, a clever but rather queer engineer from Ticino, who was a colleague in the Swiss Patent Office in Berne. A man with Latin manners – he spoke better French than German and could talk endlessly – he was capable of showing great patience in discussion. He came to be called the 'sounding-board' for his friend Albert.

Although there is no doubt that Einstein conceived his theories purely by himself, the form in which he chose to communicate them to the scientific world was considerably influenced by the endless discussions he had with Besso, and this shows that, even in the case of the most original scientific geniuses, discussion with others is invaluable.

Whitrow: As I have already mentioned, Einstein had two sons by his first wife, Mileva. The first, Hans, was born in 1904, and the second, Edward, in 1910. Edward's health broke down while he was still a boy and later he developed a mild form of schizophrenia. He died recently in Switzerland. Hans emigrated to the United States before the Second World War and is now a professor of hydraulics in the University of

California at Berkeley. Recently he was interviewed by Bernard Mayes on his youthful recollections of his father.

H. A. Einstein: One of my earliest memories goes back to when I must have been around three years old, maybe four, when my father made me a little cable car out of matchboxes. I remember that that was one of the nicest toys I had at the time and it worked. Out of just a little string and matchboxes and so on, he could make the most beautiful things. As a matter of fact, he always liked to improvise things of that sort, just as he would also like to improvise in his work in a way: for instance, when he had to give a talk he never knew ahead of time exactly what he was going to say. It would depend on the impression he got from the audience in which way he would express himself and into how much detail he would go. And so this improvisation was a very important part of his character and of his way of working. In other respects he had a character more like that of an artist than of a scientist as we usually think of them. For instance, the highest praise for a good theory or a good piece of work was not that it was correct nor that it was exact but that it was beautiful.

Mayes: What was he like as a father around the house? Did he take any interest in home life?

H. A. Einstein: Oh yes. I remember very well a time when he was still splitting wood for the stove and carrying coals up to heat in winter, because that was the way it was done at the beginning of the century. And, although he was not particularly clever with his hands to do more delicate things, he was always willing to help.

Mayes: And your mother, how did she treat his rising fame?

H. A. Einstein: She was proud of him, but that is as far as it went. It was very hard to understand, because she originally had studied with him and had been a scientist herself. But, somehow or other, with the marriage she gave up practically all of her ambitions in that direction.

Mayes: Was this the reason for their separation eventually?

H. A. Einstein: I doubt it. Why the separation came is something that was never quite clear to me. Trying to reconstruct it all afterwards, particularly from some of his own utterances, it seems that he had the impression that the family was taking a bit too much of his time, and that he had the duty to concentrate completely on his work. Personally, I do not believe that he ever achieved that, because in the family he actually had more time than when he had to look after himself and fight all the outside world alone.

Mayes: And how did your mother take the separation?

H. A. Einstein: Very hard.

Mayes: And you yourself, did it affect you?

H. A. Einstein: Yes. For quite a while, for a number of years, the separation was just a *de facto* separation and not formalized in any way. And that was probably the worst time, because then nobody knew what the future would bring – whether this was just a temporary condition or whether the marriage would end finally. It was particularly hard because it was during the war. Naturally, when everybody knew what was going to happen, then one could adjust to it.

Mayes: You say that at first the separation was *de facto.* When was that?

H. A. Einstein: In 1914.

Mayes: Did your mother communicate with your father subsequently?

H. A. Einstein: Oh yes.

Mayes: Did he feel rather cold towards her?

H. A. Einstein: I have the impression that he never felt what you call cold towards her. I mean, they may have had certain differences that I do not know of and do not understand and could not discuss, but they always communicated in a very personal and in a rather warm way.

Mayes: What about your birthday, did he remember that?

H. A. Einstein: No, he never did.

Mayes: He never remembered your birthday?

H. A. Einstein: He never did. For instance, when I was fifty years old I got a very nice letter from him, but the very first sentence was, 'Unfortunately, I have to admit that I didn't think about it, but your wife wrote me.'

Mayes: Tell me about your early days when you were a happy family together. Did you go on vacations together? Did he take that much interest in you?

H. A. Einstein: Yes. I do not remember any very extended vacation, but we often took small trips together and sometimes longer ones. In those days we were a happy family and there was nothing to indicate the separation to come.

Mayes: What sort of places did you go to?

H. A. Einstein: He was very fond of nature. He did not care for large, impressive mountains, but he liked surroundings that were gentle and colourful and gave one lightness of spirit. He needed this kind of relaxation from his intense work.

Mayes: And he was able to play with his children?

H. A. Einstein: We would play together with my toys, but he was also trying to educate us in a wider sense than the education one gets in school. He often told me that one of the most important things in his life was music. Whenever he felt that he had come to the end of the road or into a difficult situation in his work he would take refuge in music and that would usually resolve all his difficulties.

Mayes: Did he take part in disciplining you if you did things that merited it?

H. A. Einstein: Oh yes, when he felt it was necessary. And every once in a while he felt it was. I think I was quite a rascal.

Mayes: What did he do?

H. A. Einstein: Oh, he beat me up, just like anyone else would do.

Mayes: What did he beat you with? With a cudgel or something?

H. A. Einstein: Oh, I don't remember, but he did anyway.

Mayes: When he realized that he had discovered something tremendous, how did he react to this?

H. A. Einstein: Oh, just like a child. He was happy. He was walking around and telling everybody and all of a sudden would start whistling, you know. But he was always extremely careful about his findings. That means he would never accept anything until he had tested it all the way through in all directions.

Mayes: How was your relationship after you were grown up and worked independently?

H. A. Einstein: Very cordial and friendly. With both of us rather busy, the occasions of getting together were not frequent, but whenever we met we mutually reported on all the interesting developments in our fields and in our work. It may be somewhat astonishing that a theoretically-oriented mind as that of Albert Einstein would be interested in technical matters. But he thoroughly enjoyed learning about clever inventions and solutions, as he had always loved to solve certain types of puzzles. Maybe both, inventions and puzzles, reminded him of the happy, carefree and successful days at the patent office in Berne, the days before the first world war and all that followed.

Whitrow: We have seen that, as a young man in Berne, Einstein often discussed famous philosophical works with his friends. Few philosophers, however, have made any serious attempt to study Einstein's work. One who has is Bertrand Russell. Recently we asked him if he thought that Einstein was a scientist whom philosophers should study.

Russell: Einstein's stature as a scientist was, and remains, very high. He removed the mystery from gravitation which everybody since Newton had accepted with a reluctant feeling that it was unintelligible. If Einstein's reputation has appeared to diminish, that is only because recent work in physics has been mainly concerned with quantum theory. I do not think that the work of our century in either relativity or quantum theory has had any very good influence upon philosophy, but I regard this as the fault of the philosophers, who, for the most part, have not thought it necessary to master modern physics. I hope that an increasing proportion of philosophers will, as time goes on, become aware that ignorance of physics condemns any philosophy to futility.

Whitrow: I am afraid that Bertrand Russell's criticism is still true of all too many philosophers. There are, however, some exceptions; one of them is Sir Karl Popper. I asked him to tell us something about the influence that Einstein has had on his own philosophy.

Popper: Einstein's influence on my thinking has been immense. I might even say that what I have done is mainly to make explicit certain points which are implicit in the work of Einstein. I will try to sum up in four points what I have learned from Einstein directly and indirectly:

(1) Even the best-established scientific theory, such as Newton's theory of gravitation or Fresnel's theory of light, may be overthrown, or corrected, as Einstein has shown. Consequently, even the best-established scientific theory always remains a hypothesis, a conjecture.

(2) The recognition of this fact can be and should be of outstanding importance for one's own scientific work. It certainly was so for Einstein's work. He was never satisfied with any of the theories he proposed. He always tried to probe into the weak spots in order to find their limitations. And he did find them: again and again did he criticize his own work in his papers. For example, he began his famous paper of 1915 in which he first proposed the field equations for gravitation with the statement that some of his previous papers were utterly mistaken; and similarly he wrote in 1918, while replying to some criticism, that he had so far failed to distinguish between two different principles, and that his failure had led to confusion.

(3) This attitude, which may be called the critical attitude, is characteristic of the best scientific activity.

(4) With Einstein's work it became very clear that this attitude of criticism was in science something fundamentally different from what philosophers consider and describe as the 'critical attitude', or the 'sceptical attitude', or the 'attitude of doubt'.

Whitrow: Could you elaborate the difference between the critical attitude of scientists and of philosophers?

Popper: Yes. When philosophers speak of criticism they have in mind something like this. A philosopher, say Mr Adam, proposes a philosophical theory and tries to give arguments which would prove it or justify the claim that it is a true theory. Thereupon another philosopher, Mr Baker, analyses Mr Adam's proof and shows that it is invalid. Mr Baker's destructive analysis of the claims of Mr Adam to have established his theory is what philosophers usually have in mind when they speak of criticism. Or to put it another way: philosophers usually mean by criticism an analysis that aims at showing the invalidity of some arguments which have been offered in justification of the claim that a certain theory is true.

Now, it seems to have been rarely recognized that criticism in science has a very different aim and character. It is not an attack upon the proof or the justification of a scientific theory, but an attack upon the theory itself; not an attack on the claim that the theory can be *shown* to be true, but an attack on what the theory itself tells us – on its content or its *consequences*. This is so because, especially since Einstein, scientists do not seriously hold that their theories can be true or 'verified'. Nowadays they will hardly claim more than that one theory can explain more facts than other known theories, or the same facts better; that it can be tested at least as well as these other theories or even better; and that it stands up to these tests at least as well as these other theories.

This attitude became particularly clear in the case of

Einstein's criticism of Newton. Newton, in fact, had claimed that his laws of motion were not conjectural but true descriptions (if not explanations) of the facts, and that they were established by induction. But Einstein, who was a great admirer of Newton, did not criticize this mistaken claim. He did something more important; he revolutionized physics by producing an alternative to Newton's theory which not only passed all the tests which Newton's theory had passed, but also certain tests which it had failed to pass, and a few further tests which altogether went beyond the range of application of Newton's theory of gravitation. Nevertheless, Einstein regarded his own theory of gravitation merely as a step towards a better theory. Thus he wrote about his own field equations of gravitation that, as a matter of course, he never thought for a moment that his formulation of the field equations was more than a makeshift, designed to present provisionally the general principle of relativity in a concise form. And at the end of his last work, published in 1955, when discussing the pros and cons of the final results of his 35 years' search for a generalized relativity theory of a unified continuous field, he wrote that one could give good reasons showing that, and why, reality cannot be at all be represented by a continuous field.

Whitrow: Could you now tell us how this critical attitude of Einstein's which you have described has influenced your own work?

Popper: The Einsteinian revolution has influenced my own views deeply: I feel that I would never have arrived at them without him. In my view it is fundamental to science that it consists of theories which are tentative, or hypothetical, or conjectural. This means that any theory may be overthrown, however successful it may have been, and however well it may have been tested. There can be no theory more spectacularly successful than Newton's; but Einstein showed that even Newton's theory was only a conjecture. Thus, what Einstein's example may teach the philosopher is that science consists of bold speculative guesses controlled by merciless criticism which includes experimental tests.

One point about Einstein which impressed me perhaps more than any other was this: Einstein was highly critical of his own theories, not only in the sense that he was trying to discover and point out their limitations, but also in the sense that he tried, with respect to every theory he proposed, to find under what conditions he would regard it as refuted by experiment. That is, he tried to derive from each theory predictions, testable by future experiments, which he regarded as crucial for his theory, so that if his predictions were refuted he would give up the proposed theory. Thus while he regarded all physical theories – not only Newton's but also his own – as tentative guesses which might always be superseded by better ones, and which therefore could never be verified, he made it clear that he found it most important to specify the conditions which would make him look at his own theories as refuted or as falsified. This attitude became the basis of my own thesis of the logical asymmetry between verification and falsification or refutation: of the thesis that theories cannot be verified, but that they can be falsified.

Following Einstein's example, I tried at once to find out the limitations of this doctrine, and I was able to show how it was always possible to evade a refutation. But I also showed that the possibility of such an evasion did not destroy the thesis of the logical asymmetry between verification and falsification. And I pointed out that the readiness to eschew such evasions and to accept falsification was one of the basic characteristics of the critical or scientific attitude.

Whitrow: Could you give us an illustration?

Popper: Yes. I may perhaps illustrate this point by an example from Einstein's own career. When D. C. Miller, who had always been an opponent of Einstein, announced that he had overwhelming experimental evidence against special relativity, Einstein at once declared that if these results should be substantiated he would give up his theory. At the time some tests, regarded by Einstein as potential refutations, had yielded favourable results, and for this and other reasons many physicists were doubtful about Miller's alleged refutations. Moreover, Miller's results were regarded as quantitively implausible. They were, one might say, neither here nor there. Yet Einstein did not try to hedge. He made it quite clear that, if Miller's results were confirmed, he would give up special relativity and, with it, general relativity also.

This readiness to give up one's theory in accordance with the verdict of experiments is most characteristic of Einstein. It characterizes not only his critical or scientific attitude, but what may be described as his scientific realism. Although he knew that it was always possible to uphold one's theoretical constructions against unfavourable experimental evidence, he was not interested in doing so. He believed in some objectively existing reality which he tried 'to catch in a wildly speculative way', to use his own words: he was not content to find some equations fitting the observations, but he tried to grasp, to understand, this reality behind the phenomena. Yet he would have found this wild attempt uninteresting unless he could submit it to the discipline of rigorous experimental tests.

This attitude of Einstein is even today far from being generally accepted. Physicists and philosophers still speak of the verification of predictions, and even of the experimental verification of theories. But experiments have always to be interpreted in the light of theories, and theories can never be verified but remain always conjectures, wild attempts to grasp, or to understand, the hidden reality behind the phenomenal world.

Einstein's own views on the philosophy of science changed considerably during the course of his life. In his earlier writings there are many traces of positivist and conventionalist ideas. Especially noticeable is the influence of Ernst Mach, and also that of the great mathematician Henri Poincaré, who was, indeed, one of the fathers of the special theory of relativity. Einstein said things which contributed much to the positivistic doctrines of 'operational definitions' and 'meaning analysis' – doctrines that were largely based on his own famous analysis of simultaneity. In his later years, however, Einstein turned away from positivism and he told me that he regretted having given encouragement to an attitude that he now regarded not only as mistaken but as dangerous for the future development of both physical science and its philosophy. He saw more and more clearly that the growth of knowledge consisted in the formulation of theories which were far removed from observational experience. I admit, of course, that we attempt to control the purely speculative elements of our theories by ingenious experiments. Nevertheless, all our experiments are guided by theory and they cannot be interpreted except by theory. It is our inventiveness, our imagination, our intellect, and especially the use of our critical faculties in discussing and comparing our theories that make it possible for our knowledge to grow.

Whitrow: I mentioned earlier that Einstein's earliest publications were on statistical physics. An expert in this field who has a special link with Einstein is Dr R. Furth, who formerly occupied the same chair in the University of Prague that had at one time been held by Einstein. More recently Dr Furth was Reader in Theoretical Physics at Birkbeck College, in the University of London. In 1922 he assembled in book form and edited the original German edition of Einstein's papers on Brownian movement. Dr Furth stresses the importance of Einstein's contributions to statistical physics.

Furth: Most people interested in science do know of Einstein's theory of relativity and perhaps also of his theory of light-quanta or photons. But his contributions to statistical physics are, in fact, of no less importance. Statistical mechanics is a branch of theoretical physics which aims at deriving the laws for the macroscopic behaviour of material bodies from the properties of their constituent particles by a combination of causal mechanical and statistical considerations.

One of Einstein's earliest publications was a fundamental paper on the principles of statistical mechanics in which he came to conclusions that were similar to those arrived at by Gibbs, who is usually regarded as the father of statistical mechanics, but independently of Gibbs and by a different approach. In the same paper Einstein also derived a formula for the entropy of a system of atoms in terms of what nowadays is usually called the partition function. This formula is still the basis for most of the theoretical work on the thermo-dynamic properties of material bodies.

It follows from the statistical character of the macroscopic physical laws that systems which are small enough must exhibit irregular fluctuations of their parameters (for instance density, temperature, electric potential, etc.) and that it ought to be possible to observe these fluctuations with sufficiently sensitive instruments. In two papers, published in 1905 and 1906, Einstein laid the foundation of the method for calculating the magnitude of these fluctuations. In particular he came to the conclusion that small, microscopically visible particles, suspended in a liquid, should perform an irregular movement for which he derived the statistical laws. He expressed the hope that, if these could be verified experimentally, they would not only provide the first decisive proof of the existence of molecules (which at that time were still considered to be hypothetical), but would also allow the number of molecules in a given mass of a material body to be determined. Independently of Einstein, and almost at the same time, the Polish physicist Smoluchowski came to the same conclusions.

Subsequent work on this so-called 'Brownian movement' has completely confirmed these predictions, and it has become apparent that the Brownian movement, inherent in all measuring devices, sets a natural limit to the accuracy with which any kind of measurement can be carried out. Einstein further predicted that the electric charges in a circuit must perform a similar Brownian movement. This phenomenon, now usually called 'electrical noise', has in recent times become of very great practical importance, because it sets a lower limit to the strength of regular 'signals' for communication purposes if they are to be recognizable and not to be drowned in the sea of irregular 'noise'.

Later Einstein recognized that the phenomenon of spontaneous fluctuations was not restricted to material bodies but ought to be exhibited by radiation as well. At the famous Solvay Congress at Brussels in 1911 he gave a lecture in which he showed that, in order to obtain the right formula for the magnitude of these fluctuations, it was necessary to ascribe to light and other electromagnetic radiations not only wave properties but also certain particle properties, as implied by his previously conceived photon theory.

Whitrow: A well-known physicist whose opinion we have sought on Einstein's place in the history of physics is Professor Otto Frisch of the Cavendish Laboratory, Cambridge. He was asked what he considered Einstein's most important work and what he thought of Einstein as a man.

Frisch: Well, I certainly do think that his paper called 'The electrodynamics of moving bodies', which was the beginning of special relativity, was his most important work, because it knocked down the concept of absolute time and thereby opened a whole new degree of freedom for physics, and that is something which, although in a sense it was in the air, I do not think anybody else could have done. Most of the other great results of Einstein might have been produced within a few years by someone else, but I think in this case that nothing but the extraordinary power and concentration of Einstein would have been enough. The clue that led to special relativity was one of the great break-throughs comparable with the achievement of Galileo and Newton – something that only happens once in a few hundred years.

Although I often saw Einstein, I met him only briefly. I was introduced to him while I was carrying a pack of books under each arm. Einstein stood and patiently held out his hand until I had reorganized myself. The quality that dominated his personality was a very great and genuine modesty. When anybody contradicted him he thought it over and if he found he was wrong he was delighted, because he felt that he had escaped from an error and that now he knew better than before. For the same reason he never hesitated to change his opinion when he found that he had made a mistake and to say so. Indeed, there was an occasion when somebody accused him of saying something different from what he had said a few weeks previously, and Einstein replied, 'Of what concern is it to the dear Lord what I said three weeks ago?' It was just a way of saying that it did not matter. It was wrong, and now he knew better.

Whitrow: Epoch-making though Einstein's work on special relativity is – and as the years have passed since 1905 we have come to appreciate this more and more – it was nevertheless only the beginning of a long line of research that he was to pursue until the day of his death half a century later. Next we shall survey Einstein's middle period when his general theory of relativity brought him to the pinnacle of world fame.

II

THE YEARS OF FAME

1906-1932

Whitrow: Einstein's famous papers of 1905 were all written while he was still an official at the Patent Office in Berne. In 1907 he decided to apply for a post in theoretical physics in the University of Berne. He submitted as his thesis the paper on special relativity that he had published two years previously. The faculty declared that this was inadequate, pointing out, moreover, that the regulations insisted upon a hand-written thesis! For a time Einstein abandoned all thought of an academic career, but in the winter of 1908-9 he was at last allowed to give some lectures in the university to supplement those of the professor of theoretical physics. A year later he received his first full-time academic appointment as an associate professor in the University of Zürich. The salary was rather low and in March 1911 he accepted the post of professor of theoretical physics in the German University at Prague, but he was not entirely happy there. He remained in Prague barely eighteen months before returning to Zürich on being offered a chair in the Federal Institute of Technology, at which he had been a student.

For the development of his scientific work during these and the following years his friendship with his former fellow student Marcel Grossmann, who was now a professor of mathematics in the Federal Institute, was of the greatest importance. Grossmann convinced Einstein that mathematics is an essential basis of modern physics and he played a vital role in drawing Einstein's attention to the kind of mathematics required for the extension of the theory of relativity to gravitation.

Einstein was not destined, however, to remain for long in

the congenial atmosphere of Zürich. For by this time he was regarded as the rising star of theoretical physics, and at the end of 1913 he was invited to settle in Berlin as a member of the Royal Prussian Academy of Sciences and of the Kaiser Wilhelm Society, which had recently been founded as a centre of research institutes. He was also offered a chair in the University of Berlin, with no obligations attached, so that he could concentrate on research. The invitation came as a result of a petition for his election by leading German physicists, including Max Planck, the originator of the quantum theory. It is ironical that, although in this petition Einstein was praised for his theory of relativity, he was excused for having 'missed the target' in his hypothesis of light quanta, or photons!

On moving to Berlin, Einstein separated from his first wife, Mileva, and after their divorce in 1919 he married his cousin Elsa. Nevertheless, he remained on friendly terms with Mileva and their two sons. When he was awarded the Nobel Prize for Physics for 1921, he sent the money to her. Strangely enough, the prize was awarded for his work 'on the photo-electric effect and the field of theoretical physics', without mentioning relativity specifically. The electors decided to play safe, as relativity had given rise to some controversy. The Nobel lecture that Einstein delivered in July 1923, however, was devoted to that subject. Only a few years previously he had introduced his general theory of relativity. An account of this will be given by Dr D. W. Sciama.

Sciama: Einstein was twenty-six years old when he published his special theory of relativity in 1905. At that early age he had already revolutionized our understanding of space, time, and motion. Yet he remained dissatisfied. The special theory was limited in one very important respect, and in the attempt to remove this limitation Einstein created the *general* theory of relativity, perhaps the most original scientific conception ever formed by the mind of a single man.

What worried Einstein was that in special relativity only

certain types of motion are relative and not absolute. As Dr Whitrow has explained, uniform motion in a straight line has no absolute meaning; what we measure is always uniform motion relative to some other object. But everyday experience tells us that this is not true of non-uniform motion, that is, of acceleration or rotation. Anyone who has been seasick can testify to the fact that acceleration can be measured! Or, speaking more impersonally, we can say, for example, that the Earth is rotating in an absolute sense because we can detect this rotation without looking at other objects, that is, the stars. All we have to do is to note that the Earth is flattened at the poles and bulges out at the equator, unambiguous evidence of rotation.

This contrast between uniform and non-uniform motion had already been a source of puzzlement to Newton himself. A close reading of the early pages of his *Principia* suggests that he was very disturbed by it. He came to the conclusion that space has absolute properties, but that these are revealed only when a body moves non-uniformly through space.

In Newtonian language the Earth's shape is distorted by the action on it of centrifugal forces, but these forces are exerted by absolute space rather than by other physical objects, as would be the case for other types of force. We must then say that absolute space does not exert forces on bodies moving uniformly through it, so that such uniform motion cannot be detected.

It is important to realize that this particular aspect of Newtonian theory is preserved intact in special relativity. The progress of physics in the intervening years led some people to speak of the ether rather than of absolute space, but otherwise there was exactly the same contrast between uniform and non-uniform motion. This was displeasing to Einstein, who felt that the harmony of his theory of relativity required that all motion should be equally relative.

In tackling this problem Einstein made contact with a line

of thought which stretches back to the philosopher Bishop Berkeley, who wrote about absolute and relative motion not long after the publication of Newton's *Principia*. Like Einstein, Berkeley thought that all motion was relative. He tried to give a physical origin to centrifugal forces by attributing them to the action of the *stars*. According to Berkeley, as the stars revolve around the Earth they exert forces which stationary stars would not. It is these forces which flatten the poles and bulge the equator. On this view, instead of speaking of rotation relative to absolute space, we should speak of rotation relative to the stars, that is, to the bulk of the matter in the universe.

This remarkable proposal was received with scorn by Berkeley's contemporaries. The great mathematician Euler, for instance, referred to the alleged influence of the stars as 'very strange and contrary to the dogmas of metaphysics'. A hundred and fifty years later the idea was revived by the Viennese physicist and philosopher Ernst Mach, in the course of a detailed and profound assessment of Newtonian dynamics. Some of Mach's contemporaries were just as rude as Berkeley's. The young Bertrand Russell, for instance, asserted that the alleged influence of the stars 'savours of astrology and is scientifically incredible.'

Einstein, however, was much influenced by Mach's writings, and coined the term 'Mach's Principle' for the idea that distant objects determine the local state of zero acceleration, and so play a fundamental role in local dynamics. Where he went beyond his predecessors was in his detailed study of *how* distant objects can influence the local behaviour of matter. Einstein asked himself the question, what is the mechanism by which the rotating system of stars exerts centrifugal forces on the Earth? It was at this point that Einstein's genius came into play. He brought together two very simple facts, each of which was completely familiar to every physicist, but which had never before been connected together.

The first fact is that centrifugal forces depend only on the *mass* of the body on which they act, and not on any of its other properties. This is a quite trivial consequence of elementary dynamics. The second fact is that *gravitational* forces also depend only on the mass of the body on which they act, and not on any of its other properties. Nevertheless, it had to wait for Einstein to make the suggestion that centrifugal forces are actually gravitational forces exerted by the rotating system of stars. Such is the simplicity of genius.

Einstein was fond of illustrating his idea in terms of the experiences of a man confined to a closed box. If the box is at rest on the surface of the Earth the man himself will experience the pull of gravity. He will also find that when he releases objects of different mass they fall to the ground at the same rate. However, the man would have these same experiences in a completely different situation. Suppose he were actually far away from the Earth's gravity, but the box were being accelerated by a rocket. To an outside observer the objects that are released are unaccelerated, as no forces are acting on them, and it is the man who is accelerating away from them. Relative to the man, it is the objects that are accelerating. As in the gravitational case they all fall at the same rate, since this is just equal and opposite to the man's rate of acceleration. Thus the man cannot tell whether he is at rest on the Earth's surface or is in outer space being pulled by a rocket.

Now, you might think that the man could tell the difference if he studied other aspects of the behaviour of matter in his box. Perhaps some subtle atomic phenomenon would show the difference, or perhaps the behaviour of light. Einstein made the basic assumption that in every physical respect the two situations are the same. This powerful assumption has become known at the 'Principle of Equivalence'. At first sight this principle seems very audacious, including as it does all the ways in which matter and light can behave. But it finds a ready explanation in terms of Einstein's idea that accelerating stars act gravitationally on local bodies. So far we have considered the stellar system rotating relative to the Earth, but the idea applies equally to stars accelerating in a straight line. So when the man is being accelerated by a rocket he would see the stars accelerating past him. They then act on him gravitationally, so it is not really surprising if all physical processes in his neighbourhood go on in the same way as they do when the Earth's gravity acts. One gravitational field is just like another.

In a similar way if the box were falling freely under gravity, the man might think that he was in outer space with no forces acting on him at all. When he is actually falling, the man finds that the Earth's gravity is exactly cancelled by the gravitational effects of the stars accelerating past *him*. In other words, no forces are acting on him in either case, so it is not surprising that he cannot tell the difference. This point is a familiar one today, when orbiting astronauts have to face the problems of weightlessness.

With this Principle of Equivalence, we have reached the state of Einstein's thinking in 1907, just two years after the publication of his special theory of relativity. It now took him a further eight years, to 1915, to achieve a complete theory of gravitation which incorporated the Principle of Equivalence. This delay raises two questions: Why was Newton's theory of gravitation itself inadequate, and why did it take so long to replace it? It is inadequate first of all because it is inconsistent with special relativity. According to Newton, gravitational effects are propagated instantaneously. This had been questioned before Einstein by Laplace, who explicitly raised the question of the speed of propagation of gravitational effects. Special relativity makes it clear that signals cannot travel faster than light, so Newton's theory must be modified accordingly.

Now, this can be done in a straightforward manner which, however, does not of itself lead to a connection between centrifugal forces and gravitation. The reason is that the centrifugal and other forces that act on accelerating bodies have a structure more complicated than the Newtonian gravitational force, even when this force is made consistent with special relativity. For instance, when a body moves relative to the rotating earth it is acted on by the so-called Coriolis force as well as by centrifugal force. But, according to Newton, the gravitational force acting on a body does not depend on how the body moves. So a further modification of Newton's theory is needed if Einstein's explanation of Mach's principle is to be valid.

What held Einstein up all those years was the mathematical problems involved in achieving this further modification. The mathematics needed was not familiar to physicists in the early nineteen hundreds, and Einstein turned to his mathematical friend Marcel Grossman of Zürich for help. By developing an idea proposed by Minkowski in 1907, and by exploiting the so-called tensor calculus which we associate with the names of Riemann, Christoffel, and Ricci, Einstein was at last able to give his ideas perfect mathematical form in a paper published in 1916 in the *Annalen der Physik* under the title, 'Die Grundlage der Allgemeinen Relativitätstheorie' (The Foundation of the General Theory of Relativity).

In this paper Einstein proposed a new law of gravitation, the counterpart of Newton's inverse square law, but adapted in a natural way to his own mathematical description of the gravitational field. This law, enshrined in Einstein's field equations, is the one generally accepted today. Although the conceptual basis of Einstein's theory is quite different from that of Newton's, their detailed predictions turn out to be very similar so far as the solar system is concerned. The only non-Newtonian effects that have so far been measured were all mentioned by Einstein in his original paper. One is an anomaly in the motion of the planet Mercury around the Sun that was already known in the nineteenth century. This anomaly was completely cleared up by Einstein's theory.

The other two non-Newtonian effects both involved the

motion of light in a gravitational field. When light leaves a massive body it loses energy in climbing out of the body's gravitational field. This loss of energy shows up as a change in colour: the light becomes reddened, and so we speak of the 'Einstein red shift'. Attempts to detect this red shift in the light of the sun and various stars have not been too successful, but about five years ago Pound and Rebka of Harvard succeeded in detecting the colour change which occurs when light falls from a seventy-four-foot tower to the floor. In this case the light gains energy by falling and becomes bluer, but the colour change is fantastically small for such a short pathlength. By exploiting a new technique, called the Mössbauer effect, Pound and Rebka were able to measure the fractional change in frequency, which is a thousand million times smaller than the effect expected in light coming from the Sun.

The third non-Newtonian effect proposed by Einstein was the bending of light from a distant star as it passes close to the Sun. It is possible to derive an effect of this sort on the basis of Newtonian theory, but the resulting change in the apparent direction of the star is twice as great in Einstein's theory as it is in Newton's. An observational test of Einstein's prediction is not possible in the ordinary way, because when the Sun is in the sky its light is so bright that we cannot see any stars at all. However, Einstein pointed out that during a total eclipse of the Sun the stars become visible, and it should be possible to compare their positions then with their positions some months later when the same stars would be visible at night. The first attempt to test Einstein's prediction was made by British expeditions led by Sir Arthur Eddington just at the end of the Great War. At that time it was considered very striking that a group of British astronomers should attempt to test a theory proposed by a German. The results of the test were claimed as a triumph for Einstein, and his world-wide fame dates from this event. Ironically enough, attempts made at later eclipses suggest that Eddington underestimated the uncertainties inherent in such a difficult observation, and even today Einstein's prediction has not been tested with the precision one would wish.

In 1917, Einstein published a paper with the title, 'Kosmologische Betrachtungen der Allgemeinen Relativitätstheorie' (Cosmological Considerations on the General Theory of Relativity). This pioneering paper laid the foundations of modern theoretical work on the structure and history of the universe as a whole.

Einstein's aim in investigating the universe as a whole was not simply to apply his theory to a situation which obviously merited study. He had been troubled by a physical problem which had not been satisfactorily cleared up in his 1916 theory. According to that theory centrifugal forces are actually exerted by the rotating system of stars. In the hands of Berkeley and Mach this was a qualitative idea, but once Einstein set up his field equations it became quantitative. For these equations tell us the strength of the gravitational force exerted by a rotating star. The question then clearly arises: are there enough stars in the world so that their total force is equal to the observed centrifugal force? This is the question Einstein set out to answer.

Detailed calculations showed that the most important stars are very distant ones and not, as one might at first expect, the near-by ones. The reason is that there are far more distant stars, and their great number more than compensates for their great distance. The problem therefore concerns the amount of matter in the universe as a whole. Now in 1917 the expansion of the universe was not known; indeed it was not settled whether our own Galaxy, the Milky Way, comprised the total contents of the universe, or whether the so-called nebulae were galaxies like the Milky Way and lying outside it. For these reasons Einstein began by supposing that the stars are distributed more or less uniformly throughout the whole of infinite space, and are nearly at rest relative to one another.

This static universe, however, does not lead to sensible results. In particular, the total gravitational effect of all the stars on any material body in it is infinite. Einstein resolved this difficulty by showing that the curvature of space could be so great as to close the universe up so that it would only have a finite extension. In this case the total number of stars would also be finite, and so would their gravitational effect.

This remarkable model of the universe has one serious theoretical defect. It is not consistent with Einstein's field equations unless they are modified by the introduction of an extra term, a term which involved an arbitrary undetermined constant, the so-called cosmical constant. However, the need for this term was called into question a few years later by a Russian scientist, Friedmann, who discovered that expanding models of the universe were consistent with Einstein's original field equations and had no problem of infinite gravitational fields. When a few years later still the expansion of the universe was discovered observationally, Einstein immediately accepted Friedmann's models, and withdrew the cosmical constant. Nevertheless, some cosmologists prefer to keep it and attempt to determine its value directly from observation.

Of course, this observed expansion of the universe has shown that Einstein's static model is wrong, but this model possesses one feature that is still of great importance. For it leads to a definite numerical value for the amount of matter in the universe. It does so because it satisfies the condition that there should be exactly the right distribution of stars to account for the observed strength of centrifugal forces. A value for the average density of matter is also given by the expanding models. This predicted density comes fairly close to the observed density, and recent developments in astronomy suggest that a much more precise comparison will soon be possible. It seems to me that this is the most significant test of Einstein's theory, and if it is successful it will confirm the link between centrifugal-type forces and gravitation which was Einstein's greatest contribution to our understanding of nature in his general theory of relativity.

Whitrow: Although the general theory of relativity is due to Einstein alone, he made use, as Dr Sciama said, of an idea due to his former mathematics professor, Hermann Minkowski, who reformulated the special theory of relativity in terms of a new universal concept called space-time. Einstein generalized this idea mathematically and in so doing found it necessary to employ the geometrical concept of the curvature of space. By introducing sophisticated mathematical concepts like this into physics, Einstein not only abandoned the popular principle attributed to Rutherford that 'an alleged scientific discovery has no merit unless it can be explained to a barmaid', but he even outraged many professional scientists. Professor H. Levy, who was a research student at the University of Göttingen just before the First World War, recalls an occasion when Einstein came to lecture there on relativity.

Levy: I was a student at Göttingen from 1912 to 1914 and at that time it was a tremendous centre of scientific activity. Men like Klein, Hilbert, Max Born, and Richard Courant, who made great advances in pure and applied mathematics, were there. I remember Einstein coming to Göttingen to lecture on relativity. To me it was an exceptionally interesting occasion because I was greatly concerned about the distinction between a logically compelling argument and a convincing demonstration by experiment. I was puzzled by the problem of whether something which is logically compelling is also physically true. And so when Einstein came and gave his lecture I watched him very carefully to see how he expressed his ideas. In particular, I was interested in his idea of space; we had been inclined to think of space as mere emptiness between bodies, but this seemed like making nothing into something. Similarly, I realized that time is abstracted from change and that the fundamental thing we come across in the universe is change. From this we derive the ideas of space and time. To me it was exciting to hear Einstein say that he was going to take change as fundamental, and that his basic concept would be a particular combination of space and time called space-time.

I remember watching the engineering professors who were present and who were, of course, horrified by his approach, because to them reality was the wheels in machinery – really solid entities. And here was this man talking in abstract terms about space-time and the geometry of space-time, not the geometry of a surface which you can think of as a physical surface, but the geometry of space-time, and the curvature of space-time; and showing how you could explain gravitation by the way in which a body moves in space-time along a geodesic – namely the shortest curve in space-time. This was all so abstract that it became unreal to them. I remember seeing one of the professors getting up and walking out in a rage, and as he went out I heard him say, 'Das ist absolut Blödsinn' (That is absolute nonsense). Well, that really reflected the attitude of most of the engineers of the time, and they were honest in the sense that this was the way in which it appeared to them. There were others, of course, who simply thought that here was a very clever mathematician talking and after all you can expect anything from a mathematician!

Whitrow: As was mentioned earlier by Dr Sciama, the decisive event that made Einstein world famous was the success of the British eclipse expeditions of May 1919 in confirming the prediction of general relativity that the light from a remote star passing close to the Sun before reaching the Earth is deflected by a definite amount. This result was made public at a joint session of the Royal Society and the Royal Astronomical Society on 6 November 1919. The occasion was later described by the philosopher and mathematician A. N. Whitehead in a famous passage:

'It was my good fortune to be present at the meeting when

43

the Astronomer Royal for England announced that the photo-graphic plates of the famous eclipse, as measured by his colleagues in Greenwich Observatory, had verified the pre-diction of Einstein that rays of light are bent as they pass in the neighbourhood of the sun. The whole atmosphere of tense interest was exactly that of the Greek drama. We were the chorus commenting on the decree of destiny as disclosed in the development of a supreme incident. There was a dramatic quality in the very staging – the traditional ceremonial, and in the background the picture of Newton to remind us that the greatest of scientific generalizations was now, after more than two centuries, to receive its first modification.'

According to the report in *The Times* of the following day, after the President of the Royal Society had stated that they had just listened to 'one of the most momentous pronounce-ments of human thought', Sir Oliver Lodge, whose contribu-tion to the discussion had been eagerly expected, left the meeting!

Just over eighteen months later Einstein paid his first visit to this country. One of those who heard him lecture then was Professor Herbert Dingle.

Dingle: I had my first glimpse of Einstein in June 1921 when during the general excitement following the reported confirmation of his general relativity theory he came to England and lectured at King's College, London. I doubt if any scientific advance, not excluding the space explorations of more recent times, has ever roused the general public to such a pitch of enthusiasm as that which was then experienced. The idea that our most elementary notions of space and time had been found erroneous caught the imagination of the public, and 'Space Caught Bending' became the most prominent headline in a leading newspaper.

I had shortly before been a pupil of A. N. Whitehead, who had ideas of his own on the subject, and I had written an article on Einstein's theory for the *Westminster Gazette*, then one of several London evening newspapers, which appeared on the day of his visit. In recognition of this, I received a Press ticket for the lecture. It was delivered in German and my consequent inability to follow it allowed me to take undisturbed advantage of my favourable situation to observe the phenomenon who had, as it then appeared, displaced Newton as the unique interpreter of the fundamental laws of the universe. No one could live up to the expectations that the atmosphere of the time had engendered. Instead one saw a human being, to all appearances modest and serene – not the unfolder of secrets hidden from the beginning of the world but the initiator of a natural development in the progress of human understanding. It was a salutary corrective.

Whitrow: Einstein's popular fame as the creator of the theory of relativity has tended to overshadow his other contributions to physics. In fact, in the year following his main paper on general relativity he made a major contribution to the quantum theory of radiation. The significance of this paper is stressed by Professor W. H. McCrea.

McCrea: It is important to appreciate that Einstein played a leading part, possibly the leading part, in the whole revolution in physical thought that occurred in about the first quarter of our century. His work in areas other than relativity was every bit as epoch-making as his work in relativity itself. Einstein was awarded the Nobel Prize for Physics for 1921. This was ostensibly for his work on Brownian movement and on the quantum theory of radiation. Einstein's work in these fields abundantly merited this recognition, whether or not the electors took any note of his stupendous contribution in relativity theory as well. Einstein's supreme greatness was in transforming physical thinking from that of the culmination of classical physics about 1900 to that of quantum mechanics starting about 1925. In particular, far more than anyone else, he caused physicists to think in terms of probabilities. He began to do this in his early work on thermodynamics, and he brought such thinking to its first great fruition in 1905 in his work on Brownian movement and in his first work on radiation, in which he introduced the concept of light quanta or photons. Its second, even greater, fruition was in his famous paper on the quantum theory of radiation in 1917.

That paper illustrated methods that have been in use almost without change ever since, even though the majority of the users have no knowledge that it was Einstein who propounded them. It was in this paper that Einstein postulated the various transition probabilities between two states of a quantized system. It is scarcely an exaggeration to say that quantum theory has existed ever since precisely for the purpose of evaluating these probabilities. In this paper of 1917 Einstein postulated in particular the process known as stimulated emission, and inferred the properties of this process. This is the process employed in the maser – that is micro-wave amplification of stimulated emission of radiation. The light maser or laser has enabled man to shine a light on the moon and get back a detectable response, and to burn a thread-like hole through a diamond. This was dormant in Einstein's work for forty years before physicists appreciated its possibilities. Einstein's work may contain other treasures that are still hidden from us.

Einstein went on from this work of 1917 until it was possible to say, as Max Born did long afterwards, that Einstein was clearly involved in the foundation of wave mechanics and no alibi can disprove it. That refers to his work about 1924 or 1925. All that has happened in quantum theory since then is the harvest of Einstein's labours. Nevertheless, from then, again to quote Max Born, he kept himself aloof and sceptical. Perhaps the greatness of the result was even more than Einstein could bring himself to realize.

Whitrow: Professor McCrea's point concerning Einstein's role in the development of wave-mechanics is amplified by Dr R. H. Furth, who, like Professor McCrea, believes that

Einstein's contributions to quantum and statistical physics were no less important than his work in relativity.

Furth: Professor McCrea has already mentioned Einstein's statistical theory of the emission and absorption of radiation by atoms. In 1924 Einstein made a further contribution to statistical mechanics that again proved to be of prime importance to our fundamental concepts. It was initiated by a paper on a new statistical treatment of radiation by the Indian physicist Bose in which radiation was regarded as a kind of gas made up of photons. Einstein recognized that this treatment could be extended to ordinary gases consisting of atoms if one assumed that material particles, just like photons, had simultaneously wave and particle properties. This daring idea had not long before been conceived and published by the French physicist de Broglie, in a paper which at the time was not very well known. The new statistical treatment is now generally called 'Bose-Einstein statistics'. Einstein's publication in turn stimulated Schrödinger to develop his 'wave mechanics,' which is the most widely used mathematical scheme for the quantum-mechanical treatment of problems of atomic physics.

Strangely enough, although Einstein had contributed so much to the development of statistical methods in physics, he was to the end strongly opposed to the current idea, first propounded by Max Born, that the fundamental quantum laws of physics themselves were statistical in character. He was of the opinion that to use statistical methods in theoretical physics was only a useful mathematical device for dealing with phenomena involving large numbers of elementary processes, but that the basic laws for the latter were strictly causal. In a famous letter to Max Born he rejected the opposite view by saying that he 'could not believe in a dice-throwing God'.

Whitrow: Did you know Einstein personally?

Furth: I did, although I just missed being his student. He was professor of theoretical physics in the former German

University of Prague from 1910 to 1912 and I became an undergraduate there just after he had left to take up his appointment as professor in the Zürich Polytechnic. But I saw him quite frequently after the First World War at conferences and meetings and also privately. I remember one occasion at a meeting of the German Physical Society when he gave a lecture on some topic connected with the general theory of relativity and had an argument with Professor Lenard, one of the greatest opponents of relativity.

What struck me most about Einstein was his physical insight and also his simplicity and sincerity in all personal matters. When I visited him in his Berlin flat in the late twenties to invite him to speak at a meeting of the German Physical Society to be held in Prague, he received me in his study dressed as usual in an old suit and wearing sandals. Unfortunately, he could not accept the invitation, because a few months previously he had strained his heart in attempting to row a heavy sailing-boat ashore. He actually thought that he would never again be able to give a public lecture.

He was very approachable by students and indeed by anyone interested in physics who wanted to see him. I heard that when he was at Prague he told students that, whenever they had any difficulties with problems, they should come and see him and that he did not mind being interrupted in his studies, because he could resume them afterwards.

Whitrow: Another scientist who got to know Einstein personally during the nineteen-twenties was Professor Cornelius Lanczos, who is now at the Dublin Institute for Advanced Studies. I asked Professor Lanczos in a recorded interview when he first got to know Einstein.

Lanczos: I had my first discussion with Einstein when I was in Germany in 1921. I had a chance to talk to him for a few minutes between two lectures. Of course, I already had a tremendous admiration for him in my student days, and so I was in a great palpitation of heart when this opportunity came.

At that time I was very enthusiastic, because I had found a way of successive approximations for his famous gravitational equations. And this is what I tried to communicate to him, a method by which through successive mathematical approximations one is able to solve Einstein's gravitational equations, and I was, I must confess, rather proud of that. It was therefore quite a shock when he said, 'But why should anybody be interested in getting exact solutions of such an ephemeral set of equations?' I remember very well this word 'ephemeral'. It meant that he did not consider his gravitational equations as the last word. Although at the time this was quite a shock to me, later on I realized that it was a marvellous example of his complete lack of dogmatism, even in relation to his own work, and I had a tremendous admiration for this intellectual honesty.

Whitrow: Am I right in thinking that, in fact, you got to know him much better about 1928?

Lanczos: Yes, that is right. It was through the intervention of Szilard, the Hungarian physicist who at that time was also a kind of assistant to Einstein. Previously I had been corresponding with Einstein in connection with some questions of the dynamics in the de Sitter world-model. In 1928 his assistant Grommer left him and so he was looking for somebody who could help him with the mathematical development of his theory. He was always glad to find people who could work out the details. Because he was too much wedded to general ideas and general schemes, he had neither the time nor the inclination to work out the details, which he considered necessary, of course, to corroborate the conclusions of the theory. At that time I was very much interested in the equations of motion in general relativity. It was a sideline of his investigations, so he thought it might be a good occasion for us to work together. I was awarded a fellowship from the Deutsche Notgemeinschaft to enable me to spend one year in Berlin with Einstein. It was, of course, a tremendous experience, perhaps the greatest experience of my life, to see Einstein almost daily and have discussions with him of both a scientific and a non-scientific nature.

Whitrow: Would you say that in his later years he became more of a mathematician in his general philosophy of physics than when he was young?

Lanczos: I have the feeling that the young Einstein was somewhat suspicious of mathematics and did not consider it as a really constructive element of thought in physics. His attitude changed very much under the influence of general relativity, when he came to realize that in order to penetrate the depths you have to do a great deal of mathematics. Indeed, in his later years he became too much involved in mathematical speculations concerning physical reality. The turning-point in his thought was his acceptance of the principle of general covariance. In a little-known paper written with Grossman in 1911 he had already formulated the field equations of general relativity, but he felt that these equations could not be physically correct, because they would allow infinitely many solutions and the boundary conditions would not determine a unique solution. It took him three years to realize his error and return to the standpoint of general covariance.

Whitrow: Could you tell us something about Einstein's religious views?

Lanczos: There is a lovely anecdote that in some discussion between religious leaders in New York the question came up whether Einstein was a theist or an atheist and so they sent a cable to Einstein to find out. His answer was, 'I believe in the God of Spinoza who is identical with the mathematical order of the universe. I do not believe in a God who cares for the well-being and the moral doings of human beings', or something to that effect. In other words, he did not believe in a personal God, but instead he believed in a God that is the intellectual order of the universe.

Whitrow: What was Einstein's state of health when you

worked with him in Berlin? I believe that in 1926, when Einstein was awarded the gold medal of the Royal Astronomical Society, it was not possible for Jeans, the President, to present the medal to him in person, because Einstein could not make the journey to London. Was it true that he was ill for a while about this time?

Lanczos: I remember that when I came to him in 1928 he was still under the influence of some serious illness. I think he was suffering from gall-bladder trouble, but he was recovering. He was in a convalescent state and so he had to be very careful, particularly with his diet.

Whitrow: Dr E. H. Hutten of London University also has some recollections of Einstein in Berlin in the late nineteen-twenties.

Hutten: I was a student at the Berlin University when I met Einstein. After the First World War, during the nineteen-twenties, the Wednesday afternoon colloquium at the Berlin University Institute of Physics provided the occasion of presenting the most recent results of research. This colloquium was almost the only place where Einstein could be heard regularly. He disliked routine teaching and never gave the lecture courses usual in the University. He liked, however, to participate in scientific meetings. What I remember most about Einstein at this colloquium during the late 1920s was that he always regularly intervened in a quite unselfconscious manner.

Often the lecturers, sometimes very good scientists and occasionally post-graduate students, were somewhat obscure. Einstein would rise after the lecture and ask whether he might put a question. He would then go to the blackboard and begin to explain in simple terms what the lecturer had been talking about. 'I wasn't quite certain I understood you correctly,' he would say with great gentleness and then he would make clear what the lecturer had been unable to convey. Einstein always began with the simplest possible ideas and then, by describing how he saw the problem, he put it into the appropriate context. This intuitive approach was almost like painting a picture. It was an experience that taught me the difference between knowledge and understanding.

Whitrow: Hans Einstein has told us that in his opinion his father's character was more that of an artist than that of a scientist. Would you agree that Einstein had what we call the artistic temperament?

Hutten: Yes, I think in some ways he had. For him, thinking and feeling were closely allied. Feeling came first, and the thought and verbalization came later.

Whitrow: Shall I put it like this? He was more of an artist than a scholar; in other words, he did not clutter up his mind too much with other people's ideas.

Hutten: That is certainly true. In fact, I remember a beautiful remark of his when he criticized a well-known American physicist. Einstein said he 'couldn't really understand how anybody could know so much and understand so little'! Einstein always emphasized that you could know too many facts and get lost among them. Nevertheless, there existed no field of physics about which he could not immediately speak without hesitation. It did not matter whether it was a fashionable part of physics or some almost forgotten part, his listeners felt that he had the whole of physics spread out before his eyes. And yet I am quite sure that Einstein never realized what an exceptional man of genius he was. There was something childlike about him and there are many stories about his social naïvety. Presumably that was the reason why he was such an easy prey for somewhat unscrupulous people who were always importuning him and demanding his help for one cause or another.

Whitrow: Much has been written about Einstein's views on quantum mechanics and causality. I believe that, in a conversation you had with him some years after he left Europe, you had an excellent opportunity to hear what he really thought.

Hutten: Although he had contributed so much to quantum theory, Einstein was *au fond* a classical physicist and always felt himself to be so. Quantum mechanics allows only statistical laws, whereas in classical physics it is assumed, at least in principle, that events can be described by deterministic laws. Einstein felt that a statistical law leaves gaps in the description of nature, since it refers to large numbers of events rather than to individual events as does a deterministic law. Instead, he believed that the concept of field, which he had brought to such perfection in his own theory of relativity, suffices for describing physical reality.

I remember an occasion when I went to see Einstein and discuss with him his objections against quantum mechanics, which I did not share and about which I had written an essay. He read the essay and then said that he thought the arguments I put forward were logically valid and theoretically justified. It was simply a feeling on his part: he did not *like* the statistical formulation of quantum mechanics. He insisted, however, that the uncertainty principle – which forms the basis of the statistical formulation – would always remain the fundamental law of the quantum domain. Einstein's views about the lawfulness of nature were therefore not simply deterministic.

Einstein's arguments were typical of his intuitive approach. He always spoke quite openly of the aesthetic appeal, of the beauty and harmony, of certain conceptions of classical physics. It was this feeling, closely allied to his considerable musical talent, that guided him in his scientific thinking.

Whitrow: Dr Hutten, I believe that you lived near Einstein in Berlin.

Hutten: Yes, I did. I lived just a few blocks away in the same part of Berlin.

Whitrow: Did you ever meet him socially?

Hutten: Not until I was a student, although he had a slight social acquaintance with my parents through music. Sometimes my mother and he played in the same quartet.

Whitrow: In his recent autobiography *Focus and Diversions* Mr L. L. Whyte has referred to two long talks he had with Einstein in Berlin in 1929. I asked Mr Whyte to describe the circumstances in which he came to have these talks with Einstein in Berlin.

Whyte: I had a grant from the Rockefeller Foundation to study theoretical physics and I chose to go to Berlin because Einstein was there. As a matter of fact, I was too shy to press myself on Einstein in any way and I had been there for three or four months without having seen him. Then one evening I met Emil Ludwig at a party and he asked me 'What are you doing here?' I told him that I was studying theoretical physics and he at once asked me if I had met Einstein. I said 'No', and he said, 'Why not?' I told him that I was too shy and that my ideas were not mature enough to discuss with Einstein. He told me not to bother and that he would write to Einstein and that Einstein would send me an invitation that I would not like to refuse. True enough, about three days later I received a wonderful letter from Einstein saying, in German: 'Dear Mr Whyte, I hear from my friend Emil Ludwig that we both ride the same hobby horse. I always like to talk with people interested in the same things. Please ring me up and come and see me. Don't be put off by Frau Einstein. She's there to protect me.'

During the first of my talks with Einstein an amusing incident occurred. I was very nervous and still very shy and after we had been talking for about twenty minutes the maid came in with a huge bowl of soup. I wondered what was happening and I thought that this was probably a signal for me to leave. But when the girl left the room Einstein said to me in a conspiratorial whisper: 'That's a trick. If I am bored talking to somebody, when the maid comes in I don't push the bowl of soup away and the girl takes whomever I am with away and I am free.' Einstein pushed the bowl away, and so I was quite happy and much flattered and more at my ease for the rest of the talk.

Whitrow: What did you discuss with Einstein on this occasion?

Whyte: One of the reasons why I had been so shy was that I had got hold of the idea that irreversibility and processes that go one way in time ought to play a greater part in fundamental physical theory than they did at that time. I put this idea to Einstein and we discussed it and he was very friendly about it, although he left me in no doubt that he was still looking for the kind of field equations that remained invariant under time-reversal. He did, however, agree with me in one respect that I think is interesting. He believed that space-time, in which space and time are merged together so that they cannot really be sharply distinguished, is only a device used in a branch of fundamental physical theory and probably does not apply in other realms. Einstein agreed with me that in biology, for example, and in memory and in human affairs generally temporal relations can be distinguished from spatial relations. So there was a certain two-sidedness about Einstein's mind. As a theoretical physicist he was quite clear that the space-time concept was fundamental, but as a human being interested in other things besides physics he saw that temporal relations were really quite distinct from spatial relations.

Whitrow: I asked Mr Whyte what was his general impression of Einstein's position as a public figure in Germany in the late nineteen-twenties?

Whyte: Well, I remember vividly, indeed I noted it in my journal at the time, that I was deeply distressed to find that somebody of the greatness and world reputation of Einstein had already become a symbol for anti-Semitism, and that his very existence in Berlin University I thought was wrong and dangerous, because it was quite impossible for him to treat German students and foreigners in the same way. The average German student who came to him would be touched with German nationalism and so Einstein tended to be more generous to foreigners than he could possibly be to the people around him. I felt this in his treatment of me, for example. I was really embarrassed about it and I remember discussing it with some senior figure in Berlin University, who told me that it was quite impossible for Einstein then – and I am speaking of 1928 to 1929 – to fail to be conscious of the fact that he was already perhaps the dominant symbol for anti-Semitism in Germany, so that it was really very uncomfortable for him to remain there. Indeed, he left Germany for good three years later, in 1932.

Whitrow : Mr Whyte was in Berlin while Cornelius Lanczos was working with Einstein. In his autobiography Mr Whyte recounts a poignant and charming story concerning them.

Whyte : Yes, this is one of the most beautiful stories I have ever heard. It was told me within a week or two of it having happened, by Cornelius Lanczos, and since I have already put it into print I am sure he will not mind my repeating it here. In the summer of 1928 Einstein was looking for a mathematical assistant and someone put him in touch with Lanczos, who was then in Frankfurt-am-Main. He was a specialist in the mathematics of relativity theory, a Jew, and intensely devoted to Einstein's personality. Indeed, he was the ideal collaborator for Einstein. So they had a meeting – they liked one another, and the arrangement was made that they would start work together that October. The scheme was that Lanczos would go to him once a week and Einstein would present him with some task which he would then take away and work on.

Well, on the first occasion – this must have been in the early days of October 1928 – Einstein put before Lanczos a new type of wave equation or field equation and asked Lanczos to see if he could find a solution which should have certain properties. I will call them alpha, beta, and gamma. Lanczos understood the problem perfectly, was intensely proud of being given such a task by Einstein, and went away feeling very humble and unsure. But he studied the equation and after three or four days – flash! – there came into his mind the perfect solution. It had all the three properties that Einstein had asked for. Now, to Lanczos this was really something very extraordinary, because he was a very humble, religious type of personality and he felt that this inspiration had come from the heavens to him; and that he should be able to bring to Einstein in his first week in Berlin such a success moved him deeply. But he felt very humble about it, and when he reached Einstein – shall we say on Thursday afternoon – he said, 'Yes, I have been able to find a solution.' He showed it to Einstein and he demonstrated that it had the required three properties. Einstein looked at him and said, 'Yes, very interesting, quite remarkable.' There was a short silence, and then he exclaimed, rather impatiently, 'But don't you see, I gave you the wrong equation. It was quite wrong!' There was a silence. These two highly intelligent and sensitive men did not need to say anything, for they knew what a terrible thing had happened.

What actually followed was that Einstein went and fetched his violin, Cornelius Lanczos went to the piano, and they played Bach together for the rest of the hour. This story I heard from Lanczos within a week or two of it having happened.

Whitrow: Einstein's violin playing is recalled by Miss Margaret Deneke, whose house he visited several times when he came to give the Rhodes lectures in Oxford in the spring of 1931. She was interviewed recently by Christopher Sykes.

Sykes: I would like to begin by asking you, Miss Deneke, when you first met Einstein.

Miss Deneke: In 1931 when he came to Oxford. We knew of his great interest in music and we had great artists playing quartets and enjoying music in our house and they were people whose names were known to him and he was delighted to be invited to join. So he came to us. Afterwards we discovered that outsiders were not supposed to invite him, but he had found his way here and he chose to go on coming. We used to borrow instruments for him. He did not bring his violin. He played trios and quartets. He did not lead, and he always preferred to be the second violin. Now and again he would lose his place or they had to repeat something so that he came in at the right moment, and when he signed my album, which I can show to you, he would write 'Albert Einstein peccavi' because he had made a mistake. Of course, it was difficult to judge what he might have been on a better violin, because the instruments we could borrow for him were usually not of the very highest class, whereas the others all had Cremona masterpieces. Of course against them his tone seemed even smaller than it might have in an amateur quartet, because he was playing with professional musicians here.

Sykes: Can you tell me who his favourite composers were? Did he have any special favourites?

Miss Deneke: Mozart and Haydn. He would be a little hesitant about the more complicated Beethoven, but it was the classical school that appealed to him. I do not think he ever played anything very modern when he was here.

Whitrow: Another person who has a vivid recollection of Einstein in Oxford is Sir Roy Harrod.

Harrod: It was a great thrill when I first saw Einstein, the greatest living scientist and one of the greatest scientists of all time, standing there with his halo of white hair in front of the blazing log fire in Christ Church Hall, where we used to assemble before dining at high table. I soon found that my natural trepidation was quite unnecessary, as he proved to be a most easy, friendly, unassuming, and indeed lovable person. He came over to Oxford first to give the Rhodes lectures, which were a very honourable set of lectures, very well paid and we made him a member of our high table. He had rooms in Christ Church and lived with us during the period he was giving these lectures. We were so taken with him that naturally we thought it would be a great pleasure to have him again. We knew he was not being very well paid in his job in Berlin, and so we made him an offer that if he came to us every summer we would pay him a certain stipend, and it was on that basis that he came the following summer.

When we elected him we made him a member of our governing body with all the associated rights and privileges and he could have stayed another four years with us.

At the high table and in the common room he talked unaffectedly about whatever might be a matter of interest. Sometimes, probing him on political matters, I found that his views were liberal and enlightened, but not particularly deep. He evidently accepted what was in current circulation without bringing that great brain of his much to bear on such topics. In this respect he was in contrast with my colleague the late Lord Cherwell, a far less distinguished physicist. Einstein had a great regard for Lord Cherwell, whose views on political questions, although often perverse and unacceptable, invariably bore evidence of his having thought very deeply about them. But Einstein's were rather more ordinary. One might say, there was a touch of naïvety.

Whitrow: Oxford was not the only foreign university to honour Einstein by appointing him to a visiting post in the early nineteen-thirties. He was also a visiting professor at the California Institute of Technology and was there three successive winters. One who recalls meeting him during the winter of 1932–3 is Professor Dingle.

Dingle: I met him only once, if that word may be used to cover a period of a couple of months during which we were both resident at the Athenaeum of the California Institute of Technology at Pasadena. I had gone there for the session 1932–3 with a Rockefeller Fellowship to study relativity. Owing to the demands that would have been made on the time and attention of a man of Einstein's eminence, had he been freely available, the authorities of the Athenaeum allotted to him and his wife a special suite of rooms in which their meals were served, so that they did not appear in the general refectory or lounge. The seminars which he conducted, or at which he was expected to be present, were not announced, and were held in places not readily to be guessed at, so that only those whose interests were closely allied to his were able to attend.

My wife and Frau Einstein formed an acquaintance and so I obtained a back-door invitation, so to speak, to visit him in his quarters. Our talk was mostly concerned with some not very well informed ideas on the universe that had come into my mind, and he took great pains to explain to me why he did not think they would lead to any knowledge of importance. He must have feared that he had not properly expressed himself concerning the scientific point that we had been discussing, for the next morning my astonishment was great when he walked into the general refectory, the only time, I believe, that he ever appeared there, and came over to the table where I was breakfasting to draw for me a diagram to elucidate his meaning. I cannot say that it added to my enlightenment, but the kindness of the intention impressed me much.

He was evidently enjoying his stay in Pasadena, and wished to show us a picture in the adjoining bedroom which had excited his admiration. Frau Einstein, however, immediately sprang up in frantic alarm. 'Oh no, no, no, you cannot go zere. Ze bed is not made.' As though nothing was happening, Einstein gently rose with a smile on his lips, walked calmly towards the door, opened it and waved us in. The still small voice prevailed over the earthquake, wind and fire.

About this time a considerable earthquake actually occurred at Long Beach, not many miles from Pasadena, and the shock was distinctly felt there. I was in the office that had been allotted to me and at once hastened back to the Athenaeum to see if all was well. On the way I passed Einstein and Gutenberg, a distinguished seismologist who had come to work at Pasadena with the hope of experiencing an earthquake which he was unlikely to do in central Europe. They were standing on the campus closely examining a large sheet of paper. Only later did it transpire that what they had been studying was the plan of a sensitive new seismograph and they had been so absorbed in it that they had failed to notice the earthquake!

I think Einstein impressed all who met him with the gentleness and essential likeability of his character. His intellectual greatness, of course, came across, but of that one was assured beforehand, and it is difficult to say what impression he would have made on one who had never heard of him. 'He makes you feel like a schoolboy,' said R. H. Fowler, who also paid a brief visit at this time, and it was true, although anyone less like the conventional image of a schoolmaster it would be difficult to conceive. I never on any occasion saw him in the least degree ruffled, and there was not a trace of conceit or arrogance in his bearing towards anyone with whom I saw him, notwithstanding his established position as the outstanding scientist of his day. I got the impression of a man with an unqualified devotion to truth in scientific matters, as ready to discard his own views as those of another, if they failed to measure up to the demands of reason or experience. This, of course, is the popular idea of the typical scientist. The peculiarity of Einstein was that he conformed to it. In the mind's eye I can see his hand raised in gentle deprecation of those who nowadays blindly maintain the inviolability of early assertions of his which I believe he would unhesitatingly withdraw if he were now alive.

Whitrow: Did he have a striking personality that came over in conversation and lectures?

Dingle: He had a very genial personality; not the sort of imposing personality such as Sir Oliver Lodge had, for instance. He had a sense of humour certainly; not of the acute kind that one associates with Frenchmen but rather the heavy German type of wit and humour. He could laugh uproariously.

Whitrow : It was fortunate for Einstein that he was abroad when the Nazis came to power in Germany at the end of January 1933. For the greatest theoretical physicist of the age immediately became one of the principal targets for the forces of irrationality and intolerance then unleashed. Although for the rest of his life Einstein continued to work actively in pure science, his life was increasingly affected by what was happening in the world of politics. In the last of these three programmes we shall survey the final phase of Einstein's career, which he spent in the United States.

III

LAST YEARS
1933-1955

Whitrow: Early in 1933, following Hitler's rise to power in Germany, Einstein's writings on relativity were among the books burned publicly by the Nazis in the square before the State Opera House in Berlin. Einstein returned to Europe from California in the spring of that year, but did not go back to Germany. First of all he stayed in Belgium, but because it was feared that Nazi fanatics might murder him if he remained on the Continent he came to England. Later that year he left Europe for ever and settled in Princeton, New Jersey, where he had accepted a chair in the Institute for Advanced Study. This post was similar to the one he had held in Berlin. Meanwhile, all of his property in Germany had been confiscated, but fortunately most of his scientific correspondence was brought to America by diplomatic pouch. In 1936, Elsa, his second wife, died, and he was then looked after by his sister, Maja, his stepdaughter, Margot, and his secretary-housekeeper, Helen Dukas. In 1940 he became a citizen of the United States. Nearly all the papers for which Einstein is famous were written long before he emigrated to that country at the age of fifty-four. Nevertheless, unlike his great predecessor Newton who, at the same age, abandoned science on leaving Cambridge to become Warden of the Mint, Einstein continued to work as a theoretical physicist for the rest of his life.

Even after he retired from his post at the Institute in 1945, he continued to think and calculate much as before, until, after a short illness, he died of an aneurysm of the aorta on 18 April 1955 at the age of seventy-six.

We have invited a number of scientists and others who knew Einstein in his later years to tell us of their impressions, but first of all Professor W. B. Bonnor of London University will give a brief general survey of the main scientific problems that Einstein continued to work on after he left Europe.

Bonnor: Einstein's work in his later years is naturally divided into three parts. First, he made a number of investigations into aspects of general relativity; secondly, there was his criticism of the quantum theory; and thirdly, and by far the most important, there was his tireless search for a theory which would unify the whole of physics – a unified field theory, as it came to be called.

I want to take these three parts in order. First, the relativistic investigations. Except for a number of expository articles, Einstein did not often return to the general theory after he was forty. When he did so, he produced, as would be expected, important and intensely original work. For example, in 1937 he wrote a paper, in collaboration with Nathan Rosen (now at the Israel Institute of Technology), on the theory of gravitational waves.

I should explain that, in Newtonian theory, gravitation is supposed to propagate instantaneously and the question of waves does not arise. From Einstein's work in 1916 it was believed that, according to general relativity, gravitation propagates with the speed of light; and, moreover, that waves of gravitation can carry energy through empty space, just as electromagnetic waves do. In their 1937 paper Einstein and Rosen made this conjecture much more probable by producing some exact solutions of the field equations which had wave-like properties. Only within the last ten years have relativists extended and improved upon this paper by finding other wave-like solutions. Attempts are now being made to detect gravitational waves by experiment.

The most famous of Einstein's later work in general relativity concerned the motion of particles, and this was carried out in collaboration with Leopold Infeld, now the leader of a distinguished school of theoretical physics in Warsaw, and Banesh Hoffmann, now Professor at the City University of New York. To explain what was behind the work let me refer to Newton's theory of the solar system. This consisted of two distinct parts: the laws of motion, and the inverse square law of gravitation. Both these were postulated quite separately, and either might have been false without falsifying the other. In the early versions of general relativity two corresponding parts could be identified: the law of motion – called the geodesic law – on the one hand, and on the other the equations governing the strength of the gravitational field acting. It appeared probable in the nineteen-twenties from the work of Eddington, Levi-Civita and others that in general relativity the two parts were not distinct, and that the field equations prescribe not only the strength of the field but also the motion of the particles present. The famous paper of Einstein, Infeld and Hoffmann in 1938 proved that this is indeed so. General relativity is therefore not only a theory of gravitation but also a dynamics. This is another illustration of the great power of the theory.

Among Einstein's works on general relativity in this later period there appear some writings on cosmology. Dr Sciama has described how Einstein initiated this subject by his static model of the universe put forward in 1917. This model required an amendment to the field equations – the addition of the cosmological term. By 1929, from the work of the astronomer Hubble, it became clear that, because of the observed recession of the galaxies, the universe is not static; moreover, it was found that the simplest relativistic cosmological theories are not static either. In fact, as had been shown by the Russian meteorologist Friedmann as early as 1922, models of the expanding universe can be derived from the *original* field equations – that is, the equations without the cosmological term. This result led Einstein, in 1932, to abandon this term which, after all, he had introduced to cope with a situation no longer regarded as valid. In collaboration with the Dutch astronomer de Sitter he then produced a particularly simple expanding world-model which is now known as the Einstein-de Sitter universe.

One of the cosmological writings is interesting because it shows Einstein worried by a discrepancy between theory and observation. During the nineteen-forties it seemed that the interval of time during which the universe had been expanding (as judged by cosmological theory) was less than the age of the Earth (as determined by geological observations). Now, it is believed that at the start of the cosmic expansion such extreme conditions of temperature and pressure must have prevailed that the Earth could not then have existed. Hence the period of expansion of the Earth ought to be *less* than the time of expansion, not vice versa. This amounted to a serious discrepancy. Einstein was never one to give excessive weight to observations, but on this subject he wrote in 1946: 'Since determination of age (of the Earth) by these minerals is reliable in every respect, the cosmologic theory here presented would be disproved if it were found to contradict any such results. In this case I see no reasonable solution.' The resolution of the problem had to wait until 1952, when Walter Baade of Mount Wilson and Palomar Observatories announced that astronomical measurements of galactic distances had been subject to various systematic errors, which had resulted in an under-estimate of the period of cosmic expansion. The corrections were quite sufficient to reconcile theory and observation.

I shall now turn to the second of Einstein's later interests, quantum physics. In 1925 quantum theory took a decisive turn of which Einstein strongly disapproved. It had been clear since the work of Planck and of Einstein himself at the beginning of the century that atomic theory must bring substantial changes to classical mechanics. Thus classical mechanics was being attacked on two fronts, so to speak. On the one side there was special and general relativity attacking chiefly the ideas of space and time. On the other was the quantum theory attacking the classical conception of energy, and other concepts besides. Energy, like matter, was now to be thought of as existing in small discrete lumps, or quanta. Other physical variables, such as momentum and spin, were quantized too. These innovations were essential for Niels Bohr's explanation of atomic structure, put forward in 1913. It was widely believed that, successful as Bohr's picture was as a model for spectroscopy, it could hardly be more than a stopgap.

Einstein accepted these amendments to classical conceptions as a temporary expedient until a logical basis could be found. For most physicists the logical basis appeared during the years 1925 to 1927 in work by Werner Heisenberg and Erwin Schrödinger. This revolution Einstein could not and never did accept. It struck at the roots of classical thought, for it questioned the deterministic view on which science had hitherto been based. This view is well illustrated by Newtonian mechanics according to which, given the positions and velocities of each of a system of interacting particles at a certain instant, their motions can be calculated for all time. According to the Uncertainty Principle of the new quantum theory, however, no physicist could ever obtain precise information of the simultaneous position and velocity of a single atomic particle. Consequently there is no possibility of predicting exactly the future of the system. All that physicists could hope to predict are probabilities that it will behave in certain ways.

The assumption that our knowledge of physical systems is necessarily incomplete stuck in Einstein's throat, and he mocked the new statistical physics with the phrase, 'God does not play dice with the world'. He devised several hypothetical experiments – so-called 'thought experiments' – aimed to show either that complete information can be got in principle, provided the apparatus is constructed correctly, or that the new quantum mechanics gives an interpretation of them which is obviously absurd. The thought experiments were characteristically ingenious, and caused a good deal of head-scratching in the thirties and forties; but the consensus now is that they were unsuccessful: that they did not show quantum mechanics to be self-contradictory, or to be untenable for other reasons. Nevertheless, Einstein continued to maintain that, in spite of its many successes, statistical quantum theory was only a makeshift, and that a more fundamental and completely deterministic theory would eventually be discovered.

To show how he thought the discovery might come, let me turn to the third strand in his later thought. As I said before, the search for a unified field theory was to him the most compelling of his problems at that time. Indeed, it was probably the most compelling of his whole life and certainly the most ambitious.

To appreciate what Einstein was trying to do it is important to understand what physicists today mean by a field, and how it differs from the much simpler idea of absolute empty space, which was at the basis of Newtonian mechanics. At the simplest level, the field of a piece of matter is a region of influence round about it. Thus we speak of the gravitational field of the Earth, or the electric field of a proton. At this stage there is no conflict at all between the idea of field and that of absolute empty space. But there is really much more to a field than this. Take, for example, the electromagnetic field, such as that created by a radio transmitter. In this case energy actually travels through space from the transmitter to the receiving set. Thus at a given moment the space round the transmitter contains energy. According to special relativity, energy and matter are essentially the same thing. Can we, then, really say that the space round the transmitter is empty? The field itself begins to take on a character as real as that of the particles that generate it.

The importance of the idea of field was enhanced by the discovery of general relativity, in which the conception reaches a most beautiful and powerful realization. This expresses itself not merely in physical simplicity nor in the compactness of the mathematical formulation. The power of the conception resides in something I mentioned before – that the field equations of general relativity determine the motion of particles present. In electromagnetic theory Maxwell's field equations do not determine the motion of electric charges: to do this one has to bring in something from outside, namely Newton's laws of motion. But in general relativity the field takes on a more dominant role and even the particles must submit to it. Nevertheless, the idea of the particle is not eliminated completely. Particles remain in the theory as sources of the gravitational field, which reacts back on them and makes them move. It was Einstein's aim to abolish the particle altogether and to express all physical reality by field.

To common sense this may seem a quixotic enterprise. According to the senses, matter has a real existence quite distinct from, say, the gravitational field. The programme becomes more plausible when we consider the short-range but extremely strong field of atomic physics. The hardness of steel is due to the repulsive field of force between atoms in the metal and atoms in one's finger. Our perception of matter is thus our interaction with a field of force.

Einstein's intention was to construct the properties of matter out of properties of fields. A particle would be simply a region in which the field became very intense – rather like a knot in the uniform grain of a piece of wood. The difference between matter and its surrounding field would be of degree and not of kind. The two fundamental field theories are electromagnetism and general relativity. Einstein hoped by uniting these to get a superfield theory which would explain the whole of physical reality. He expected, if he achieved this, that the uncertainties of the quantum theory would be overcome and deterministic physics reinstated. It was a grand conception. One can easily see why it dominated the last years of Einstein's life. And not only his. Many of the greatest physicists of the last generation were fascinated by the prospect of a unified field theory.

The difficulties were formidable. Gravity, the force for which the field concept is best developed, is a very weak force which becomes perceptible only for large bodies. For charged particles, such as electrons, the electric repulsion is enormously greater than the gravitational attraction. For this reason it has not been possible to carry out any experiments showing how the electromagnetic and gravitational fields interact with one another. Thus there are no empirical facts to guide us in the unification of the fields.

More important still has been the lack of any real physical insight into the problem. As Dr Sciama said, Einstein was led to general relativity by his thought experiment about a box falling freely under gravity, which told him that a uniform gravitational field is equivalent to an acceleration. This flash of physical inspiration guided the mathematical development of general relativity. Einstein never experienced any corresponding insight into the physical basis of the unified field theory.

Instead, the inspiration had to come from the mathematics. The mathematics of relativity permits of extension in many ways, and the search for a unified field theory became the search for the correct mathematical generalization of general relativity. To more empirically minded physicists this intensely mathematical approach seemed a case of the tail wagging the dog. But in the early nineteen-twenties, with the triumphs of general relativity fresh in physicists' minds, it seemed only a matter of time before this method would reveal the ultimate reality.

The first unified field theory was produced not by Einstein but by the mathematician Hermann Weyl, as early as 1918. He generalized the geometry of Riemann, which had been used as a basis for general relativity. Eddington reformulated

Weyl's theory, and so did Einstein in his first papers on unified field theory in 1923. Nevertheless, it became clear that Weyl's theory led to no new physical results, and certainly did not apply to the microscopic world of the quantum theory.

Over the years Einstein and other mathematical physicists published many unified field theories, none of them successful. A famous one, put forward by Kaluza, recast the mathematics of relativity in a space of five dimensions instead of the customary four. Kaluza hoped, rather naïvely it now seems to us, that unification of gravitation and electromagnetism would follow by a suitable interpretation of the fifth dimension. Einstein took up Kaluza's work, but once more no new physical understanding appeared. The most celebrated unified field theory, called the generalized theory of gravitation, was published by Einstein in 1945, and independently by Schrödinger at about the same time. By then Schrödinger was disillusioned by quantum mechanics, which he himself had helped to create. For some years both these eminent men believed that the new theory would solve the problems which they thought still beset fundamental physics.

The mathematics of the theory is very complicated, and it is necessary to use approximate methods. With these it was shown that the theory failed to satisfy the basic requirement that its field equations must determine the motion of electric charges present. This was enough to convince Schrödinger that the theory was useless. Einstein was not prepared to accept the approximate results: he maintained that the theory could be judged only if certain *exact* solutions of the generalized field equations were found. These exact solutions, he believed, would represent matter by pure field. This was the position he took up until his death.

The exact solutions Einstein wanted have still not been discovered, and it is not even known whether they exist. Almost nobody now works on the generalized theory of gravitation or, indeed, on any other unified field theory. The common feeling is that no further progress is likely in this direction.

Scientists sometimes say that to close a gap can be as important as opening one. Although we may look back with some sadness on Einstein's failure to crown his life's work with a successful unified field theory, nobody who understands what he was looking for can blame him for making the attempt.

Whitrow: A possible explanation for Einstein's failure to develop a unified field theory is suggested by one of his collaborators in the nineteen-thirties, Professor Peter Bergmann, now of Syracuse University.

Bergmann: I was privileged to work under Einstein during part of the period when he was primarily concerned with unified field theory. When Einstein succeeded in 1915 in creating a new theory of gravitation it turned out that you could no longer separate the properties of space and time from the properties of the gravitational field. The geometry had become the gravitational field or, if you prefer it so, you can say that the gravitational field had become the geometry. But all the other fields of which the physicist is aware continued to play their roles against the geometric background that had been so intimately fused into the gravitational field. Einstein thought that this was a basically unsatisfactory situation; he spent the best part of the remainder of his life attempting to fuse the remaining physical fields with the gravitational field and the geometry. The number of attempts he made is quite large, but none was successful.

I believe that Einstein's work on unified field theory was ahead of its time. His own attempts were directed towards unifying electromagnetic and gravitational theory, but today we know that there are a number of forces of great importance for the constitution of elementary particles and their interactions that are neither electromagnetic nor gravitational. It seems to me that a truly unified field theory should cover these forces too. So I would say that Einstein's programme was ahead of its time, simply because we were not in possession of the body of facts needed for the next major step in the unification of physical theories. I have no doubt that this programme will have to be taken up again at its proper time, but we shall only know that it is the proper time when the programme is successful.

I think that, in choosing his own main field of work, Einstein was attracted by that part of physics which he thought was in greatest need of clarification.

It should also be mentioned that Einstein was concerned with the corrupting influence that the need to be successful has on the scientist. He frequently discussed this both in print and in conversation. He suggested that it would be a very nice profession for a scientist to be a lighthouse-keeper, for it would not be very demanding intellectually and would leave plenty of time to think about other matters. Although this suggestion was not very realistic, it is obvious that Einstein's concern for the integrity of scientists and of scientific research was not an idle worry. Today, when scientists require large amounts of money to pursue their experimental work, when they have to connive to get computer time to do their work, we find that the scientist is often obliged to direct his attention to acquiring prestige and influence so that he may have access to the means of research. Certainly many young scientists feel that they cannot afford to devote themselves to research programmes where it may take a long time to get significant results.

Whitrow: As Professor Bonnor has said, the most famous of Einstein's scientific achievements in his later years is the paper that he wrote with Infeld and Hoffmann showing that general relativity is not only a theory of gravitation but is also a dynamics or theory of motion. Professor Banesh Hoffmann tells us what it was like to work with Einstein.

Hoffmann: It was my good fortune to work with Einstein. You would imagine that this would be a wonderful opportunity to see how his mind worked and so you would learn how to become a great scientist yourself. Unfortunately, no such revelation was forthcoming. Genius simply cannot be reduced to a set of simple rules for anyone to follow. But this happened way back in 1937 – at Princeton, New Jersey, in the United States. Leopold Infeld and I had been doing some research work together and when it was finished we screwed up our courage and went to see Einstein in the hope that we could work with him. He received us very warmly and he proposed two possible problems for us to work on. We chose the one that seemed the easier, naturally, and it was lucky we did, because the other was so difficult it has not yet been solved.

Even so, the one we chose was extraordinarily complicated and on more than one occasion we almost gave up in dismay because of the difficulties that it presented. Einstein outlined the idea to us and showed us his initial calculations on it; from these calculations it seemed that there was a good chance that the idea would work. They also laid down the basic plan that was ultimately to guide the whole research, but before we were through we had many unpleasant surprises and had to make several ingenious and subtle modifications in the plan of campaign.

Soon we fell into a routine. Each morning Infeld and I would go to Einstein's house and would show Einstein our calculations and the three of us would then consider their implications and review the overall strategy in the light of what the calculations showed. Then in the afternoon Infeld and I would calculate together and also separately, so that we would have more to show Einstein at the next morning session. Sometimes things went smoothly and at other times we seemed to be up against a blank wall. Often the blank wall was obvious to Infeld and me, but occasionally when we thought all was well Einstein would point out a major conceptual difficulty that was so deep and subtle that Infeld and I had been unable to see it, and yet it was so simple that, once

Einstein had discovered it, it seemed to have been staring us in the face all the time.

Whenever we came to an impasse the three of us had heated discussions – in English for my benefit, because my German was not too fluent – but when the argument became really intricate Einstein, without realizing it, would lapse into German. He thought more readily in his native tongue. Infeld would join him in that tongue, while I struggled so hard to follow what was being said that I rarely had time to interject a remark till the excitement died down.

When it became clear, as it often did, that even resorting to German did not solve the problem, we would all pause, and then Einstein would stand up quietly and say, in his quaint English, 'I will a little tink'. So saying he would pace up and down or walk around in circles, all the time twirling a lock of his long, greying hair around his forefinger. At these moments of high drama Infeld and I would remain completely still, not daring to move or make a sound, lest we interrupt his train of thought. A minute would pass in this way and another, and Infeld and I would eye each other silently while Einstein continued pacing and all the time twirling his hair. There was a dreamy, far-away, and yet a sort of inward look on his face. There was no appearance at all of intense concentration. Another minute would pass and another, and then all of a sudden Einstein would visibly relax and a smile would light up his face. No longer did he pace and twirl his hair. He seemed to come back to his surroundings and to notice us once more, and then he would tell us the solution to the problem and almost always the solution worked.

So here we were, with the magic performed triumphantly and the solution sometimes was so simple we could have kicked ourselves for not having been able to think of it by ourselves. But that magic was performed invisibly in the recesses of Einstein's mind, by a process that we could not fathom. From this point of view the whole thing was completely frustrating.

But, from the more immediately practical point of view, it was just the opposite, since it opened a way to further progress and without it we should never have been able to bring the research to a successful conclusion.

Whitrow: A vivid impression of what Einstein was like as a lecturer is given by Professor J. A. Wheeler of Princeton. He was interviewed by Christopher Sykes.

Wheeler: As a very young man I first came to Princeton on a visit. I heard Einstein giving a seminar at the Institute for Advanced Study. The seminar was, of course, my first opportunity to see Einstein. I never had a chance on that occasion to talk to him, but what struck me more than any other single thing in his seminar was the wholesale way that he dealt in the equations. Every physicist that I know is a retail dealer in equations, but this wholesale method of counting up how many equations one has and how many unknowns one has, and then how many things are left over undetermined without taking the trouble to look at all the details, was to me a new experience. Since that time I have seen the fullness of Einstein's point of view and that was certainly one lesson well gained.

I only heard Einstein give two lectures in my life. The other one was, in fact, the last that he ever gave in this world, shortly after I had taken a group of students to see him. In all of his lectures and in his discussions he spoke what I thought was beautiful English, each word pronounced so carefully to be sure there was no accent in it. But the power of expression that he showed was to me most impressive. The words came not gushing out, but each one carefully measured and mouthed in full.

Sykes: He had a certain charming humour, hadn't he?

Wheeler: And a mordant humour, too. On the fireplace over the Professors' room at the Institute there is an inscription which Einstein contributed at the time the building was being constructed. It says, 'Raffiniert ist der Herr Gott aber boshaft ist Er nicht.' (God is cunning but He is not malicious.) In other words the world has been put together in a very complicated and subtle way, but still the Lord gives us a chance to find out how it is done.

Whitrow: Professor Bergmann reminds us that, despite the highly mathematical nature of much of his work, Einstein was essentially a physicist.

Bergmann: Einstein to the end of his life retained an earthy and practical sense which I think makes it proper to say that he remained in his attitude a physicist; that is, a natural scientist rather than a mathematician. One of the more amusing aspects of Einstein's work was a paper that he published on the so-called meandering of rivers. As probably most of you know, the course of a river once it leaves the mountains and spreads itself in the plain becomes tortuous. It is not quite clear when you look at it straightforwardly why the water, instead of proceeding on the most direct course, should instead have a tendency to bend this way and that way. It turns out – and this is Einstein's discovery – that the water in the river bed tends not only to flow forward downstream, but also to carry out a rotatory motion like a screw and it is this rotatory motion which tends to transport material, that is sand or soil, from the outside towards the inside and thus to make the curve become more curved as time goes on.

Whitrow: Another of Einstein's collaborators, who worked with him in the mid-nineteen-forties, Professor Ernst Straus, has a nice story to tell of Einstein's belief in the essential simplicity of his ideas.

Straus: He was convinced that his ideas were fundamentally very simple despite their very heavy mathematical mechanisms. He had a firm conviction, which I do not think was justified, that he could explain it to everybody. For instance, as I remember quite clearly, we were working on something in unified field theory and he came down rather cheerfully and said, 'I explained it this morning to my sister and she also thinks that it is a very good idea.' Now, his sister was a very intelligent woman, but she was a philologist and had not the slightest idea of any of his work. Since she was a very good listener, he liked to explain his newest ideas to her. He did not feel that he understood something until he himself had understood it in these simple and basic terms.

Frequently, I remember, if I brought up a mathematical argument that seemed to him unduly abstract, he would say, 'I am convicted but not convinced', that is to say, he could no longer get out of agreeing that it was correct, but he did not yet feel that he had understood why it was so. For, in order to convince himself that something was so, he had to reduce it to a certain simplicity of concept.

Whitrow: Einstein differed from most other scientists of his day in the importance that he attached to the philosophical aspects of physics. Indeed, it was for essentially philosophical reasons that he could not accept modern quantum theory as developed and interpreted by Heisenberg, Born, and Bohr. His most important criticism of that theory was made in a celebrated paper that he wrote in collaboration with Boris Podolsky and Nathan Rosen in 1935. Professor Hoffmann comments on Einstein's standpoint.

Hoffmann: Einstein simply could not bring himself to accept the quantum theory and that is strange because, after all, he was one of the most important people in the development of the theory and in bringing about its acceptance. Now, in this paper with Podolsky and Rosen, he pointed out an extremely subtle difficulty in the interpretation of the quantum theory, and indeed there are people now who still believe that this difficulty has never satisfactorily been overcome. But in those days, as now, belief in the impeccable validity of the quantum theory was the fashionable view, and in science, as in other walks of life, fashion exerts an almost overwhelming coercion. Einstein took the unfashionable view and he was well aware of this. Indeed, he made a wry comment about it that I would like to tell you. He said that, when this paper with Podolsky and Rosen was published, he received several letters from scientists eagerly pointing out to him just where the argument was wrong. What amused Einstein was that, while all the scientists were quite positive that the argument was wrong, they all gave different reasons for their belief!

Whitrow: The title of the paper was 'Can Quantum-Mechanical Description of Physical Reality be Considered Complete?' As a preliminary to understanding the point of this paper, an important feature of quantum mechanics must be described, at least briefly.

In quantum mechanics, which must be used rather than classical mechanics when we study the motion of very small objects like electrons, there is, as Professor Bonnor briefly mentioned, a peculiar restriction in principle on the accuracy of measurement that has no counterpart in classical physics. For example, in order to locate an object such as an electron with great accuracy, we must use a 'microscope' that operates not with ordinary light but with gamma-rays, that is with electromagnetic radiation of much smaller wavelength. This means that, in principle, we must expose the object to a collision with quanta of extremely short wavelength. The shorter the wavelength, however, the greater the momenta of these quanta and consequently the more violent the collision. As a result, the momentum of the electron, or other object studied, is changed by an amount that tends to increase with the accuracy attained in determining the electron's position.

Generally, in a physical system subject to quantum mechanics, the process of measuring any physical quantity of the system tends to disturb the system in an unpredictable way so that the measure of some associated quantity is changed by an unpredictable amount. Position and momentum provide one example of such a pair of associated quantities, and time and energy another.

Heisenberg was the first to analyse a number of measuring

processes in this way and was thereby led to his famous Uncertainty Principle. This says that, in the case of two associated quantities, the uncertainty in the determination of the one multiplied by the uncertainty in the determination of the other is greater than, or of the same order of magnitude as, Planck's constant. According to this principle, the more accurately one measures either quantity the greater becomes the uncertainty in the value of the other, due to the disturbance of the system by the measurement.

Professor Rosen summarizes the main ideas of the paper he published with Einstein and Podolsky in the *Physical Review* in 1935.

Rosen: The paper begins by pointing out that, in discussing a physical theory, one must distinguish between the physical, or objective, reality, which is independent of any theory, and the physical concepts of the theory which are intended to correspond with the objective reality. In judging the merits of a theory one asks two questions:

(1) Is the theory correct?
(2) Is the description given by the theory complete?

Only if affirmative answers are given to both questions may the theory be considered satisfactory.

The *correctness* of the theory is judged by the degree of agreement between the conclusions of the theory and human experience. It is the second question, as applied to quantum mechanics, that is the subject of the paper under discussion: what does one mean by the *completeness* of the theory? It is asserted that the following is a necessary requirement for a theory to be complete: every element of the physical reality must have a counterpart in the physical theory.

However, we see that we need some way of deciding what are the elements of the physical reality. For the present purpose the following criterion is taken: if, without in any way disturbing a system, we can predict with certainty the value of a physical quantity, then there exists an element of physical reality corresponding to this physical quantity.

Now, it is shown in the paper by means of an example – and this is perhaps the most important part of the paper – that the following situation can arise in quantum mechanics: two systems, denoted by I and II, interact for a certain time, after which the interaction is switched off. The systems are then left in states which are correlated in such a way that, by carrying out a certain measurement on system I, we get the value of a particular quantity P for II; or, by carrying out a different measurement on I, we get the value of a quantity Q for II, where P and Q are associated quantities, such as position and momentum.

Since at the time of the measurement the two systems are no longer interacting, the measurement on system I does not disturb system II. We see then that, without disturbing system II, we can determine, or predict, the value of either P or Q. Therefore, according to our criterion, both of these are elements of the physical reality. However, because of Heisenberg's Principle, the quantum-mechanical formalism does not allow us to describe both of them simultaneously. Hence, the quantum-mechanical description is incomplete.

This conclusion led to a considerable amount of discussion. Those opposed to this conclusion, and this seemed to include nearly all those working in quantum mechanics, questioned either the criterion of an element of reality or the phrase 'without disturbing the system' as applied to the example mentioned. One gets the impression that Einstein's opponents believed at the time that their arguments completely demolished the paper. However, it seems that its ghost continues to haunt those concerned with the foundations of quantum mechanics. The question raised thirty years ago is still being discussed.

Whitrow: Despite his great fame, Einstein always remained accessible for interviews, particularly with young scientists.

His courteous patience with them is illustrated by the experience of Professor Hermann Bondi.

Bondi: In the summer of 1947 I met Einstein for the first and only time. I was a young and quite unknown scientist then, because my real work only started after the war. And what impressed me so enormously at this interview was his kindly interest in younger people. He listened with the greatest attention to points that must have been reasonably obvious to him, possibilities of investigating a problem where he must have known every highway and byway. He was entirely patient, extremely courteous, and very interested. In those days I was still interested in the problem of unified field theories which occupied Einstein's attention very much at the time. Personally, I lost interest very soon afterwards, but at the time I was aware of some of the difficulties and problems and discussed particular proposals and suggestions with him and voiced my criticism of some other attempts. What I remember most clearly was that when I put down a suggestion that seemed to me cogent and reasonable, he did not in the least contest this, but he only said, 'Oh, how ugly.' As soon as an equation seemed to him to be ugly, he really rather lost interest in it and could not understand why somebody else was willing to spend much time on it. He was quite convinced that beauty was a guiding principle in the search for important results in theoretical physics.

Whitrow: One of the last interviews granted by Einstein only a fortnight before he died was with the historian of science Professor Bernard Cohen of Harvard. When he was in England recently I asked Professor Cohen about this interview and of the impression he formed of Einstein's mental state.

Cohen: Well, I cannot give a comparative statement, because, of course, I only saw him this one time, but I certainly had the impression of a man who was alert, who was eagerly interested in all kinds of questions, such as those of the history

and philosophy of science that we talked about, and in particular I had the feeling of a man who had a tremendous zest for
information of all kinds and who was willing to exchange ideas
and to find out what was going on in other fields.

Whitrow: Can you recall the general circumstances of your
interview?

Cohen: Dominating the whole room was a very large
window looking out on a pleasant green view. The free spaces
on the walls were occupied by portraits of the founders of
electromagnetic theory, Faraday and Maxwell. I stayed in that
room for a few minutes waiting, and then Einstein came in,
greeted me with a smile, went out of the room again and came
back filling his pipe with tobacco.

He was dressed in a very informal way: he had an open
shirt and a blue sweat-shirt over it and was wearing grey
flannel trousers and leather slippers. It was a little chilly,
being early spring, and he had a blanket that he tucked around
his feet. And I looked at his face, which was a very beautiful
and extraordinary face. It was contemplatively tragic, deeply
lined, and yet had sparkling eyes that gave him a quality of
agelessness. While we spoke his eyes watered continuously and
in moments of laughter he would have to wipe away the tears
with the back of his hand. He spoke softly but clearly. He had
a remarkable command of English, although he spoke with a
marked German accent. And the contrast, I may say, between
the soft speech and ringing laughter was enormous. He
enjoyed making jokes and every time he made a point he liked,
or heard something that appealed to him, he would burst into
booming laughter that would just echo from wall to wall. It
was most extraordinary. I had been prepared from having
seen pictures to know what he would look like and what he
would wear and I had heard the study described, but I was
totally unprepared for this roaring, booming, friendly, all-
enveloping laughter.

Whitrow: Did he speak about Newton in this interview?

Cohen: As a matter of fact, it was owing to my interest in Newton that I had the chance to see Einstein. Some time before I had published an edition of Newton's *Opticks*; and I had remarked that Newton's ideas in this field had been given a new lease of life in the twentieth century, because it seemed to many people that Newton's theory, that optical phenomena were caused by waves guiding corpuscles, was similar to some aspects of modern quantum theory. I wrote in my introduction that the importance of Newton's book was due to whatever influence it exerted on people who read it when it was published. I then added that undoubtedly Einstein had never read Newton's *Opticks* nor had he ever been influenced by it. Later I began to wonder on what authority I had made this statement. How did I know that Einstein had never read the book? I decided that the way to find out was to write and ask him. I got back an extremely friendly letter saying that he had not studied it, at least not at all profoundly. His letter showed such an awareness of historical problems that I decided to write and ask for an interview when I should be next in Princeton. That is how I came to visit him.

Whitrow: What aspects of Newton did you discuss?

Cohen: Among other things, we discussed Newton's theology, and this was very interesting, because it revealed some of Einstein's own feelings on the subject. I tried to explain to Einstein how Newton believed that there had been a primitive Christianity that has somehow been corrupted and that some primitive original message was to be found in the Scriptures hidden behind some later corruptions. He therefore took certain important words and tried to find out their meaning in different usages. I thought Einstein would be interested to see how a scientifically minded man would treat such questions. On the contrary, Einstein said that he thought that this was a great weakness in Newton and he went on to explain why. It seemed to him that, if Newton found that his ideas were at variance with orthodox ideas, he ought to have rejected orthodox views. For instance, if Newton could not agree with the accepted interpretation of the Scriptures, why did he believe that the Scriptures were true all the same? Although I tried to explain that a man's mind is imprisoned by his culture and the environment that moulds him, I did not get very far, and it did not seem to me that I ought to press the point.

I did refer, however, to the vast quantity of theological manuscripts that Newton had written and I mentioned the fact that he had never wanted to publish any of them. Einstein was very much impressed by this. He said that it indicated that Newton was fully aware that his theological conclusions were imperfect, and he thought that one should admire Newton for not wanting to present to the public anything that did not come up to his high standards. And then he said with great passion that, if Newton did not want to publish his own writings, he hoped no one else would. Obviously he was concerned with the problem of privacy because he had been so hounded by reporters and people who wanted every detail of his life all the time. He said that he felt every man has a right to privacy after his death. On the other hand, he felt that it was permissible to publish the correspondence of great men because, if you wrote a letter and sent it, you clearly intended that it should be read. But even there he added a warning that some letters are personal and should be withheld.

Whitrow: When you took your leave of Einstein did you have any feeling that this might be the last time you would see him?

Cohen: No. As a matter of fact, I think one of the most extraordinary parts of my visit to Einstein was the fact that despite his age and despite the appearance of age, as shown in his very white hair and in his eyes, which were watering while he talked, nevertheless in the way he talked about the different problems we discussed, and also in his wonderful booming laughter, I had the feeling of someone really filled with life and exuberance.

Whitrow: Although Einstein was always at heart an internationalist, the tribulations that afflicted many of his fellow Jews, particularly after 1933, made him increasingly sympathetic with the Zionist movement and later with the State of Israel. In his will he directed that, after the deaths of Margot Einstein and Helen Dukas, all his personal and scientific archives, now at Princeton, be sent to Israel. Einstein's sympathy with Zionism led to a curious incident a few years before he died. Christopher Sykes recently questioned Dr Meyer Weisgal of the Weizmann Institute in Israel about this.

Sykes: Do you know anything about the offer that Einstein should become President of the State of Israel?

Weisgal: Yes, after Dr Weizmann's death in 1952, Mr Ben Gurion naïvely, I believe, had a brainwave and asked the Israeli Ambassador in Washington to inquire from Einstein whether he would take the Presidency. If B.G. had asked me, I would have told him not to waste his time. Einstein was as far removed from the trappings, pomp, and circumstance of a presidential office as I am or you are from an understanding of the theory of relativity.

Sykes: What impression did Einstein make on you when you first met him?

Weisgal: He seemed to me a very shy man who did not really belong to our physical world, despite the fact that his genius lay in revealing to us the secrets of this physical world. He would rather probe its mysteries than confront the glare of public adulation.

Sykes: Can you give an example of this?

Weisgal: In 1921 Einstein went to the United States with Dr Weizmann on an important Zionist mission. There was a huge meeting in New York with about a hundred thousand Jews gathered in and around Madison Square. Einstein was sitting on the platform and marvelling at the great crowd and wondering how he came to be there. He was promised that he would not have to speak, but the audience insisted that he should and finally forced him to his feet. He came to the rostrum and said in German, for he did not speak English at that time: 'Weizmann ist mein Führer. Folge ihm. Ich habe gesprochen.' (Weizmann is my leader. Follow him. I have spoken.)

Sykes: Did Einstein keep up his interest in Zionism?

Weisgal: Between 1921 and 1948 he was alternately pleased and disappointed with the Zionist movement. At one time he became somewhat disenchanted with the Hebrew University in Jerusalem, because he feared that it was becoming too much of a Jewish theological seminary, but that danger has long since passed and the University is now a genuine citadel of learning. In his later years Einstein became very much attached to the Weizmann Institute. He left all his scientific papers to the Institute.

Whitrow: The question of Einstein's sympathy with Zionism was also raised by Christopher Sykes when interviewing Professor Straus.

Sykes: Is it true that, in his support of Zionism, Einstein became anti-British?

Straus: He was a Zionist on general humanitarian grounds rather than on nationalistic grounds. He felt that Zionism was the only way in which the Jewish problem in Europe could be settled even before the Second World War and certainly afterwards. His attitude was anti-British at one time only because he felt that when there was trouble in a colonial country it was due to the colonial power promoting strife. He was firmly convinced that in India there would have been no trouble between Moslems and Hindus if it had not been for the British Government. Similarly, in Palestine he believed that the quarrel between Jews and Arabs was largely due to the British. However, in 1947 he joined with President Magnus of the Hebrew University in an appeal against Jewish terror. He was never in favour of aggressive nationalism, but he felt that a Jewish homeland in Palestine was essential to save the remaining Jews in Europe and he hoped that it would be a State with a high moral tone. After the State of Israel was established he said that somehow he felt happy he was not there to be involved in the deviations from the high moral tone that he detected in news of that country.

Whitrow: In his later years Einstein became increasingly concerned with the social consequences of science. Originally, like Spinoza, who refused the offer of a university post and preferred to earn his living grinding lenses so as to be free to meditate in seclusion, Einstein hankered after an independent existence. Indeed, as Professor Bergmann mentioned, he often used to say that the ideal social position for an original thinker is to be a lighthouse-keeper! Nevertheless, as the international scene darkened, his sense of social responsibility and involvement increased, until in the summer of 1939 he had to make perhaps the most momentous decision that has ever confronted a man of science.

In July of that year the physicists Leo Szilard and E. P. Wigner approached him with the object of preventing a terrible calamity. For they felt that Einstein's name would make a far greater impression than that of anyone else in drawing the attention of the United States Government to the problem of the uranium bomb and the danger of the Nazis being allowed to develop it by laying hands on Belgian reserves of uranium. 'The possibility of a chain reaction in uranium had not occurred to Einstein,' Szilard relates, 'but, almost as soon as I began to tell him about it, he realized what the consequences might be and immediately signified his readiness to help us.' So when, on their second visit early in August, Szilard and Wigner brought Einstein a letter addressed to President Roosevelt, he agreed to sign it.

Nevertheless, this letter was very slow in making its effect, and the following March it was again necessary for Einstein to write to Roosevelt on the apparent interest of the Nazis in supplies of uranium. Five years later, when it was no longer necessary to worry about the Nazis, Einstein wrote a third time to the President, again at Szilard's suggestion, urging him to prevent the bomb from being used against Japan, but this letter was found unopened on Roosevelt's desk the day he died.

A detailed study of Einstein's views on the social influence of science and of his personal contribution to this influence has been made by Professor Eugeniusz Olszewski of the Warsaw Institute of Technology, whom I interviewed recently in Poland.

Olszewski: The atomic bomb changed the moral and social situation of science. In a letter written to Queen Elizabeth of Belgium in 1954, Einstein said, 'Strange that science, which in the old days seemed harmless, should have evolved into a nightmare that causes everyone to tremble.' Einstein, it is well known, participated in the production of the atomic bomb by one single act of signing a letter to President Roosevelt on the necessity of conducting large-scale experimentation with regard to the feasibility of producing an atom bomb. But already at the end of 1945 he said, 'Today the physicists who participated in forwarding the most formidable and dangerous weapons of all times are harrassed by a feeling of responsibility, not to say guilt.' Half a year later Einstein agreed to serve as chairman of the newly formed emergency committee of atomic scientists which had the purpose to advance the use of atomic energy in ways beneficial to mankind, to diffuse knowledge and information about atomic energy, and to promote the general understanding of its consequences. The committee was active until the end of 1948.

Whitrow: On the occasion of Einstein's seventieth birthday one of the most moving tributes that was paid to him was in a broadcast on the Third Programme by Bertrand Russell. He concluded by referring to Einstein's attempts after the war to work politically with American nuclear scientists to seek international agreement for the control of atomic energy. But this problem, as Russell wryly remarked, is more difficult than that of relativity. Bertrand Russell has specially recorded for our programme on Einstein this further tribute to his memory.

Russell: Of all the public figures that I have known, Einstein was the one who commanded my most whole-hearted admiration. I got to know him fairly well at a time when we were both at Princeton in the early forties. He arranged to have a little meeting at his house once a week at which there would be some one or two eminent physicists and myself. We used to argue about moot points in the philosophy of physics in an attempt, sometimes vain, to reach fundamental agreement. We did not, in those days, talk much about international politics, chiefly because in such matters we all thought alike. There was, however, one exceptional occasion. I remarked at a meeting that, when Germany had been defeated, the victors would lend money to the German Government and would forget the German crimes. Einstein indignantly repudiated the suggestion, but subsequent experience proved that on this occasion he was mistaken.

When, in the early fifties, the danger of nuclear war began to seem likely to cause universal ruin, I began to feel that this was a risk far greater and far more terrible than any of those with which governments were concerning themselves. I expressed my fears in a BBC broadcast on 23 December 1954. I sent the text of this to Einstein asking him whether he thought it possible that we could get scientists on both sides of the Iron Curtain to sign such a statement. He replied that he was too ill to work himself, but would gladly join me in signing any appeal on the subject that I might draw up and would suggest names of scientists to whom it might be sent.

I adapted the broadcast into a form of an appeal from scientists which I sent to certain eminent physicists including Einstein. After I had obtained a number of signatures from men of the highest scientific eminence, but not from Einstein, I learnt of his death during a flight from Rome to Paris. When I reached Paris, I found his letter agreeing to sign, dated two days before his death and the last public act of his life. This manifesto, known as the Russell-Einstein Manifesto because of the dramatic circumstances of Einstein's signing it, was the origin of the Pugwash Scientific Conferences.

Einstein was not only a great scientist, he was a great man. He stood for peace in a world drifting towards war. He remained sane in a mad world, and liberal in a world of fanatics.

Whitrow: Although Einstein was no doubt a somewhat more complex personality than generally imagined, he was essentially a man of basic goodness and general kindliness. He was endowed with a robust sense of humour that survived the afflictions of life into old age. He was completely lacking in pomposity and in that sense of self-importance that so often corrupts lesser men. He was entirely free from the trappings of convention, not only in his thought but in his way of life.

Once when a too fulsome speech was being made at a dinner in America in his honour, he whispered to his neighbour, referring to himself, 'But he doesn't wear any socks!'

The secret of his success in physics was his irrepressible curiosity and his unsurpassed ability to think clearly. With characteristic modesty he said of himself, 'God gave me the stubbornness of a mule and a fairly keen scent.' Even if in later life he seems to have lost the magic touch, his earlier achievements entitle us to regard him as the principal founder of twentieth-century theoretical physics.

BIBLIOGRAPHY

1. *Biographies*

There is no definitive biography of Einstein. A lively account of his life is given in

Carl Seelig, *Albert Einstein: a Documentary Biography*, translated by Mervyn Savill. London, Staples Press, 1956.

More attention is paid to his scientific work, but there is less biographical detail, in

Philipp Frank, *Einstein: his Life and Times*. London, Jonathan Cape, 1948.

A recent biography by a distinguished Russian authority is

B. Kuznetsov, *Einstein*, translated by V. Talmy. Moscow, Progress Publishers, 1965.

Indispensable for any serious study of Einstein are his 'Autobiographical Notes' in

P. A. Schilpp (editor), *Albert Einstein: Philosopher-Scientist*, Evanston, Illinois, The Library of Living Philosophers, Inc., 1949.

2. *Writings by Einstein*

A bibliography of Einstein's scientific papers was compiled by E. Weil and published by him in 1960. A bibliography containing both scientific and non-scientific items is given in the book edited by Schilpp, pages 694–756.

Einstein's most important original papers on relativity, edited with notes by A. Sommerfeld, are available in English translation in

A. Einstein and others, *The Principle of Relativity*, translated by W. Perrett and G. B. Jeffery. London, Methuen, 1923.

Einstein's papers on Brownian motion, edited with notes by R. Furth are contained in

Albert Einstein, *Investigations on the Theory of the Brownian Movement*, translated by A. D. Cowper. London, Methuen, 1926.

An English translation of Einstein's paper of 1905 on the photoelectric effect was recently made available in

'Einstein's Proposal of the Photon Concept – a Translation of the *Annalen der Physik* Paper of 1905' by A. B. Arons and M. B. Peppard, *American Journal of Physics*, vol. 33, No. 5, May 1965, 367–74.

Einstein's most famous popular exposition of his work in relativity is

Albert Einstein, *Relativity: the Special and the General Theory*, translated by R. W. Lawson. 15th edition. London, Methuen, 1955.

A more technical account of the subject will be found in

Albert Einstein, *The Meaning of Relativity*, translated by E. P. Adams. 4th edition. London, Methuen, 1950.

Some of Einstein's more popular occasional pieces, on both scientific and non-scientific topics are in

Albert Einstein, *The World as I see It*, translated by A. Harris. London, John Lane, 1935;

later issued in expanded form as *Ideas and Opinions*, translated by Sonja Bargmann. New York, Crown Publications, 1954.

Other essays will be found in

Albert Einstein, *Out of My Later Years*. London, Thames and Hudson, 1950.

An authoritative account of Einstein's pacificist activities and a carefully translated and edited selection of his voluminous correspondence relating to international affairs is given in

Einstein on Peace, edited by O. Nathan and H. Norden. London, Methuen, 1963.

A CATALOG OF SELECTED
DOVER BOOKS
IN ALL FIELDS OF INTEREST

A CATALOG OF SELECTED DOVER
BOOKS IN ALL FIELDS OF INTEREST

CONCERNING THE SPIRITUAL IN ART, Wassily Kandinsky. Pioneering work by father of abstract art. Thoughts on color theory, nature of art. Analysis of earlier masters. 12 illustrations. 80pp. of text. 5⅜ × 8½.   23411-8 Pa. $3.95

ANIMALS: 1,419 Copyright-Free Illustrations of Mammals, Birds, Fish, Insects, etc., Jim Harter (ed.). Clear wood engravings present, in extremely lifelike poses, over 1,000 species of animals. One of the most extensive pictorial sourcebooks of its kind. Captions. Index. 284pp. 9 × 12.   23766-4 Pa. $10.95

CELTIC ART: The Methods of Construction, George Bain. Simple geometric techniques for making Celtic interlacements, spirals, Kells-type initials, animals, humans, etc. Over 500 illustrations. 160pp. 9 × 12. (USO)   22923-8 Pa. $8.95

AN ATLAS OF ANATOMY FOR ARTISTS, Fritz Schider. Most thorough reference work on art anatomy in the world. Hundreds of illustrations, including selections from works by Vesalius, Leonardo, Goya, Ingres, Michelangelo, others. 593 illustrations. 192pp. 7⅛ × 10¼.   20241-0 Pa. $8.95

CELTIC HAND STROKE-BY-STROKE (Irish Half-Uncial from "The Book of Kells"): An Arthur Baker Calligraphy Manual, Arthur Baker. Complete guide to creating each letter of the alphabet in distinctive Celtic manner. Covers hand position, strokes, pens, inks, paper, more. Illustrated. 48pp. 8¼ × 11.
24336-2 Pa. $3.95

EASY ORIGAMI, John Montroll. Charming collection of 32 projects (hat, cup, pelican, piano, swan, many more) specially designed for the novice origami hobbyist. Clearly illustrated easy-to-follow instructions insure that even beginning papercrafters will achieve successful results. 48pp. 8¼ × 11.   27298-2 Pa. $2.95

THE COMPLETE BOOK OF BIRDHOUSE CONSTRUCTION FOR WOOD-WORKERS, Scott D. Campbell. Detailed instructions, illustrations, tables. Also data on bird habitat and instinct patterns. Bibliography. 3 tables. 63 illustrations in 15 figures. 48pp. 5¼ × 8½.   24407-5 Pa. $1.95

BLOOMINGDALE'S ILLUSTRATED 1886 CATALOG: Fashions, Dry Goods and Housewares, Bloomingdale Brothers. Famed merchants' extremely rare catalog depicting about 1,700 products: clothing, housewares, firearms, dry goods, jewelry, more. Invaluable for dating, identifying vintage items. Also, copyright-free graphics for artists, designers. Co-published with Henry Ford Museum & Green-field Village. 160pp. 8¼ × 11.   25780-0 Pa. $8.95

HISTORIC COSTUME IN PICTURES, Braun & Schneider. Over 1,450 costumed figures in clearly detailed engravings—from dawn of civilization to end of 19th century. Captions. Many folk costumes. 256pp. 8⅜ × 11¾.   23150-X Pa. $10.95

STICKLEY CRAFTSMAN FURNITURE CATALOGS, Gustav Stickley and L. & J. G. Stickley. Beautiful, functional furniture in two authentic catalogs from 1910. 594 illustrations, including 277 photos, show settles, rockers, armchairs, reclining chairs, bookcases, desks, tables. 183pp. 6½ × 9¼. 23838-5 Pa. $8.95

AMERICAN LOCOMOTIVES IN HISTORIC PHOTOGRAPHS: 1858 to 1949, Ron Ziel (ed.). A rare collection of 126 meticulously detailed official photographs, called "builder portraits," of American locomotives that majestically chronicle the rise of steam locomotive power in America. Introduction. Detailed captions. xi + 129pp. 9 × 12. 27393-8 Pa. $12.95

AMERICA'S LIGHTHOUSES: An Illustrated History, Francis Ross Holland, Jr. Delightfully written, profusely illustrated fact-filled survey of over 200 American lighthouses since 1716. History, anecdotes, technological advances, more. 240pp. 8 × 10¾. 25576-X Pa. $10.95

TOWARDS A NEW ARCHITECTURE, Le Corbusier. Pioneering manifesto by founder of "International School." Technical and aesthetic theories, views of industry, economics, relation of form to function, "mass-production split" and much more. Profusely illustrated. 320pp. 6⅛ × 9¼. (USO) 25023-7 Pa. $8.95

HOW THE OTHER HALF LIVES, Jacob Riis. Famous journalistic record, exposing poverty and degradation of New York slums around 1900, by major social reformer. 100 striking and influential photographs. 233pp. 10 × 7⅞.
22012-5 Pa $10.95

FRUIT KEY AND TWIG KEY TO TREES AND SHRUBS, William M. Harlow. One of the handiest and most widely used identification aids. Fruit key covers 120 deciduous and evergreen species; twig key 160 deciduous species. Easily used. Over 300 photographs. 126pp. 5⅜ × 8½. 20511-8 Pa. $2.95

COMMON BIRD SONGS, Dr. Donald J. Borror. Songs of 60 most common U.S. birds: robins, sparrows, cardinals, bluejays, finches, more—arranged in order of increasing complexity. Up to 9 variations of songs of each species.
Cassette and manual 99911-4 $8.95

ORCHIDS AS HOUSE PLANTS, Rebecca Tyson Northen. Grow cattleyas and many other kinds of orchids—in a window, in a case, or under artificial light. 63 illustrations. 148pp. 5⅜ × 8½. 23261-1 Pa. $3.95

MONSTER MAZES, Dave Phillips. Masterful mazes at four levels of difficulty. Avoid deadly perils and evil creatures to find magical treasures. Solutions for all 32 exciting illustrated puzzles. 48pp. 8¼ × 11. 26005-4 Pa. $2.95

MOZART'S DON GIOVANNI (DOVER OPERA LIBRETTO SERIES), Wolfgang Amadeus Mozart. Introduced and translated by Ellen H. Bleiler. Standard Italian libretto, with complete English translation. Convenient and thoroughly portable—an ideal companion for reading along with a recording or the performance itself. Introduction. List of characters. Plot summary. 121pp. 5¼ × 8½.
24944-1 Pa. $2.95

TECHNICAL MANUAL AND DICTIONARY OF CLASSICAL BALLET, Gail Grant. Defines, explains, comments on steps, movements, poses and concepts. 15-page pictorial section. Basic book for student, viewer. 127pp. 5⅜ × 8½.
21843-0 Pa. $3.95

BRASS INSTRUMENTS: Their History and Development, Anthony Baines. Authoritative, updated survey of the evolution of trumpets, trombones, bugles, cornets, French horns, tubas and other brass wind instruments. Over 140 illustrations and 48 music examples. Corrected and updated by author. New preface. Bibliography. 320pp. 5⅜ × 8½. 27574-4 Pa. $9.95

HOLLYWOOD GLAMOR PORTRAITS, John Kobal (ed.). 145 photos from 1926–49. Harlow, Gable, Bogart, Bacall; 94 stars in all. Full background on photographers, technical aspects. 160pp. 8⅜ × 11¼. 23352-9 Pa. $9.95

MAX AND MORITZ, Wilhelm Busch. Great humor classic in both German and English. Also 10 other works: "Cat and Mouse," "Plisch and Plumm," etc. 216pp. 5⅜ × 8½. 20181-3 Pa. $5.95

THE RAVEN AND OTHER FAVORITE POEMS, Edgar Allan Poe. Over 40 of the author's most memorable poems: "The Bells," "Ulalume," "Israfel," "To Helen," "The Conqueror Worm," "Eldorado," "Annabel Lee," many more. Alphabetic lists of titles and first lines. 64pp. 5³⁄₁₆ × 8¼. 26685-0 Pa. $1.00

SEVEN SCIENCE FICTION NOVELS, H. G. Wells. The standard collection of the great novels. Complete, unabridged. First Men in the Moon, Island of Dr. Moreau, War of the Worlds, Food of the Gods, Invisible Man, Time Machine, In the Days of the Comet. Total of 1,015pp. 5⅜ × 8½. (USO) 20264-X Clothbd. $29.95

AMULETS AND SUPERSTITIONS, E. A. Wallis Budge. Comprehensive discourse on origin, powers of amulets in many ancient cultures: Arab, Persian, Babylonian, Assyrian, Egyptian, Gnostic, Hebrew, Phoenician, Syriac, etc. Covers cross, swastika, crucifix, seals, rings, stones, etc. 584pp. 5⅜ × 8½. 23573-4 Pa. $10.95

RUSSIAN STORIES/PYCCKNE PACCKA3bl: A Dual-Language Book, edited by Gleb Struve. Twelve tales by such masters as Chekhov, Tolstoy, Dostoevsky, Pushkin, others. Excellent word-for-word English translations on facing pages, plus teaching and study aids, Russian/English vocabulary, biographical/critical introductions, more. 416pp. 5⅜ × 8½. 26244-8 Pa. $7.95

PHILADELPHIA THEN AND NOW: 60 Sites Photographed in the Past and Present, Kenneth Finkel and Susan Oyama. Rare photographs of City Hall, Logan Square, Independence Hall, Betsy Ross House, other landmarks juxtaposed with contemporary views. Captures changing face of historic city. Introduction. Captions. 128pp. 8¼ × 11. 25790-8 Pa. $9.95

AIA ARCHITECTURAL GUIDE TO NASSAU AND SUFFOLK COUNTIES, LONG ISLAND, The American Institute of Architects, Long Island Chapter, and the Society for the Preservation of Long Island Antiquities. Comprehensive, well-researched and generously illustrated volume brings to life over three centuries of Long Island's great architectural heritage. More than 240 photographs with authoritative, extensively detailed captions. 176pp. 8¼ × 11. 26946-9 Pa. $14.95

NORTH AMERICAN INDIAN LIFE: Customs and Traditions of 23 Tribes, Elsie Clews Parsons (ed.). 27 fictionalized essays by noted anthropologists examine religion, customs, government, additional facets of life among the Winnebago, Crow, Zuni, Eskimo, other tribes. 480pp. 6⅛ × 9¼. 27377-6 Pa. $10.95

THE BEST TALES OF HOFFMANN, E. T. A. Hoffmann. 10 of Hoffmann's most important stories: "Nutcracker and the King of Mice," "The Golden Flowerpot," etc. 458pp. 5⅜ × 8½. 21793-0 Pa. $8.95

FROM FETISH TO GOD IN ANCIENT EGYPT, E. A. Wallis Budge. Rich detailed survey of Egyptian conception of "God" and gods, magic, cult of animals, Osiris, more. Also, superb English translations of hymns and legends. 240 illustrations. 545pp. 5⅜ × 8½. 25803-3 Pa. $10.95

FRENCH STORIES/CONTES FRANÇAIS: A Dual-Language Book, Wallace Fowlie. Ten stories by French masters, Voltaire to Camus: "Micromegas" by Voltaire; "The Atheist's Mass" by Balzac; "Minuet" by de Maupassant; "The Guest" by Camus, six more. Excellent English translations on facing pages. Also French-English vocabulary list, exercises, more. 352pp. 5⅜ × 8½. 26443-2 Pa. $8.95

CHICAGO AT THE TURN OF THE CENTURY IN PHOTOGRAPHS: 122 Historic Views from the Collections of the Chicago Historical Society, Larry A. Viskochil. Rare large-format prints offer detailed views of City Hall, State Street, the Loop, Hull House, Union Station, many other landmarks, circa 1904–1913. Introduction. Captions. Maps. 144pp. 9⅜ × 12¼. 24656-6 Pa. $12.95

OLD BROOKLYN IN EARLY PHOTOGRAPHS, 1865–1929, William Lee Younger. Luna Park, Gravesend race track, construction of Grand Army Plaza, moving of Hotel Brighton, etc. 157 previously unpublished photographs. 165pp. 8⅞ × 11¾. 23587-4 Pa. $12.95

THE MYTHS OF THE NORTH AMERICAN INDIANS, Lewis Spence. Rich anthology of the myths and legends of the Algonquins, Iroquois, Pawnees and Sioux, prefaced by an extensive historical and ethnological commentary. 36 illustrations. 480pp. 5⅜ × 8½. 25967-6 Pa. $8.95

AN ENCYCLOPEDIA OF BATTLES: Accounts of Over 1,560 Battles from 1479 B.C. to the Present, David Eggenberger. Essential details of every major battle in recorded history from the first battle of Megiddo in 1479 B.C. to Grenada in 1984. List of Battle Maps. New Appendix covering the years 1967–1984. Index. 99 illustrations. 544pp. 6½ × 9¼. 24913-1 Pa. $14.95

SAILING ALONE AROUND THE WORLD, Captain Joshua Slocum. First man to sail around the world, alone, in small boat. One of great feats of seamanship told in delightful manner. 67 illustrations. 294pp. 5⅜ × 8½. 20326-3 Pa. $4.95

ANARCHISM AND OTHER ESSAYS, Emma Goldman. Powerful, penetrating, prophetic essays on direct action, role of minorities, prison reform, puritan hypocrisy, violence, etc. 271pp. 5⅜ × 8½. 22484-8 Pa. $5.95

MYTHS OF THE HINDUS AND BUDDHISTS, Ananda K. Coomaraswamy and Sister Nivedita. Great stories of the epics; deeds of Krishna, Shiva, taken from puranas, Vedas, folk tales; etc. 32 illustrations. 400pp. 5⅜ × 8½. 21759-0 Pa. $8.95

BEYOND PSYCHOLOGY, Otto Rank. Fear of death, desire of immortality, nature of sexuality, social organization, creativity, according to Rankian system. 291pp. 5⅜ × 8½. 20485-5 Pa. $7.95

A THEOLOGICO-POLITICAL TREATISE, Benedict Spinoza. Also contains unfinished Political Treatise. Great classic on religious liberty, theory of government on common consent. R. Elwes translation. Total of 421pp. 5⅜ × 8½. 20249-6 Pa. $7.95

MY BONDAGE AND MY FREEDOM, Frederick Douglass. Born a slave, Douglass became outspoken force in antislavery movement. The best of Douglass' autobiographies. Graphic description of slave life. 464pp. 5⅜ × 8½.　22457-0 Pa. $7.95

FOLLOWING THE EQUATOR: A Journey Around the World, Mark Twain. Fascinating humorous account of 1897 voyage to Hawaii, Australia, India, New Zealand, etc. Ironic, bemused reports on peoples, customs, climate, flora and fauna, politics, much more. 197 illustrations. 720pp. 5⅜ × 8½.　26113-1 Pa. $15.95

THE PEOPLE CALLED SHAKERS, Edward D. Andrews. Definitive study of Shakers: origins, beliefs, practices, dances, social organization, furniture and crafts, etc. 33 illustrations. 351pp. 5⅜ × 8½.　21081-2 Pa. $7.95

THE MYTHS OF GREECE AND ROME, H. A. Guerber. A classic of mythology, generously illustrated, long prized for its simple, graphic, accurate retelling of the principal myths of Greece and Rome, and for its commentary on their origins and significance. With 64 illustrations by Michelangelo, Raphael, Titian, Rubens, Canova, Bernini and others. 480pp. 5⅜ × 8½.　27584-1 Pa. $9.95

PSYCHOLOGY OF MUSIC, Carl E. Seashore. Classic work discusses music as a medium from psychological viewpoint. Clear treatment of physical acoustics, auditory apparatus, sound perception, development of musical skills, nature of musical feeling, host of other topics. 88 figures. 408pp. 5⅜ × 8½. 21851-1 Pa. $8.95

THE PHILOSOPHY OF HISTORY, Georg W. Hegel. Great classic of Western thought develops concept that history is not chance but rational process, the evolution of freedom. 457pp. 5⅜ × 8½.　20112-0 Pa. $8.95

THE BOOK OF TEA, Kakuzo Okakura. Minor classic of the Orient: entertaining, charming explanation, interpretation of traditional Japanese culture in terms of tea ceremony. 94pp. 5⅜ × 8½.　20070-1 Pa. $2.95

LIFE IN ANCIENT EGYPT, Adolf Erman. Fullest, most thorough, detailed older account with much not in more recent books, domestic life, religion, magic, medicine, commerce, much more. Many illustrations reproduce tomb paintings, carvings, hieroglyphs, etc. 597pp. 5⅜ × 8½.　22632-8 Pa. $9.95

SUNDIALS, Their Theory and Construction, Albert Waugh. Far and away the best, most thorough coverage of ideas, mathematics concerned, types, construction, adjusting anywhere. Simple, nontechnical treatment allows even children to build several of these dials. Over 100 illustrations. 230pp. 5⅜ × 8½.　22947-5 Pa. $5.95

DYNAMICS OF FLUIDS IN POROUS MEDIA, Jacob Bear. For advanced students of ground water hydrology, soil mechanics and physics, drainage and irrigation engineering, and more. 335 illustrations. Exercises, with answers. 784pp. 6⅛ × 9¼.　65675-6 Pa. $19.95

SONGS OF EXPERIENCE: Facsimile Reproduction with 26 Plates in Full Color, William Blake. 26 full-color plates from a rare 1826 edition. Includes "The Tyger," "London," "Holy Thursday," and other poems. Printed text of poems. 48pp. 5¼ × 7.　24636-1 Pa. $3.95

OLD-TIME VIGNETTES IN FULL COLOR, Carol Belanger Grafton (ed.). Over 390 charming, often sentimental illustrations, selected from archives of Victorian graphics—pretty women posing, children playing, food, flowers, kittens and puppies, smiling cherubs, birds and butterflies, much more. All copyright-free. 48pp. 9¼ × 12¼.　27269-9 Pa. $5.95

PERSPECTIVE FOR ARTISTS, Rex Vicat Cole. Depth, perspective of sky and sea, shadows, much more, not usually covered. 391 diagrams, 81 reproductions of drawings and paintings. 279pp. 5⅜ × 8½. 22487-2 Pa. $6.95

DRAWING THE LIVING FIGURE, Joseph Sheppard. Innovative approach to artistic anatomy focuses on specifics of surface anatomy, rather than muscles and bones. Over 170 drawings of live models in front, back and side views, and in widely varying poses. Accompanying diagrams. 177 illustrations. Introduction. Index. 144pp. 8⅜ × 11¼. 26723-7 Pa. $7.95

GOTHIC AND OLD ENGLISH ALPHABETS: 100 Complete Fonts, Dan X. Solo. Add power, elegance to posters, signs, other graphics with 100 stunning copyright-free alphabets: Blackstone, Dolbey, Germania, 97 more—including many lower-case, numerals, punctuation marks. 104pp. 8⅛ × 11. 24695-7 Pa. $6.95

HOW TO DO BEADWORK, Mary White. Fundamental book on craft from simple projects to five-bead chains and woven works. 106 illustrations. 142pp. 5⅜ × 8. 20697-1 Pa. $4.95

THE BOOK OF WOOD CARVING, Charles Marshall Sayers. Finest book for beginners discusses fundamentals and offers 34 designs. "Absolutely first rate . . . well thought out and well executed."—E. J. Tangerman. 118pp. 7¾ × 10⅝. 23654-4 Pa. $5.95

ILLUSTRATED CATALOG OF CIVIL WAR MILITARY GOODS: Union Army Weapons, Insignia, Uniform Accessories, and Other Equipment, Schuyler, Hartley, and Graham. Rare, profusely illustrated 1846 catalog includes Union Army uniform and dress regulations, arms and ammunition, coats, insignia, flags, swords, rifles, etc. 226 illustrations. 160pp. 9 × 12. 24939-5 Pa. $10.95

WOMEN'S FASHIONS OF THE EARLY 1900s: An Unabridged Republication of "New York Fashions, 1909," National Cloak & Suit Co. Rare catalog of mail-order fashions documents women's and children's clothing styles shortly after the turn of the century. Captions offer full descriptions, prices. Invaluable resource for fashion, costume historians. Approximately 725 illustrations. 128pp. 8⅜ × 11¼. 27276-1 Pa. $10.95

THE 1912 AND 1915 GUSTAV STICKLEY FURNITURE CATALOGS, Gustav Stickley. With over 200 detailed illustrations and descriptions, these two catalogs are essential reading and reference materials and identification guides for Stickley furniture. Captions cite materials, dimensions and prices. 112pp. 6½ × 9¼. 26676-1 Pa. $9.95

EARLY AMERICAN LOCOMOTIVES, John H. White, Jr. Finest locomotive engravings from early 19th century: historical (1804–74), main-line (after 1870), special, foreign, etc. 147 plates. 142pp. 11⅜ × 8¼. 22772-3 Pa. $8.95

THE TALL SHIPS OF TODAY IN PHOTOGRAPHS, Frank O. Braynard. Lavishly illustrated tribute to nearly 100 majestic contemporary sailing vessels: Amerigo Vespucci, Clearwater, Constitution, Eagle, Mayflower, Sea Cloud, Victory, many more. Authoritative captions provide statistics, background on each ship. 190 black-and-white photographs and illustrations. Introduction. 128pp. 8⅞ × 11¾. 27163-3 Pa. $12.95

EARLY NINETEENTH-CENTURY CRAFTS AND TRADES, Peter Stockham (ed.). Extremely rare 1807 volume describes to youngsters the crafts and trades of the day: brickmaker, weaver, dressmaker, bookbinder, ropemaker, saddler, many more. Quaint prose, charming illustrations for each craft. 20 black-and-white line illustrations. 192pp. 4⅝ × 6. 27293-1 Pa. $4.95

VICTORIAN FASHIONS AND COSTUMES FROM HARPER'S BAZAR, 1867–1898, Stella Blum (ed.). Day costumes, evening wear, sports clothes, shoes, hats, other accessories in over 1,000 detailed engravings. 320pp. 9⅜ × 12¼.
22990-4 Pa. $12.95

GUSTAV STICKLEY, THE CRAFTSMAN, Mary Ann Smith. Superb study surveys broad scope of Stickley's achievement, especially in architecture. Design philosophy, rise and fall of the Craftsman empire, descriptions and floor plans for many Craftsman houses, more. 86 black-and-white halftones. 31 line illustrations. Introduction. 208pp. 6½ × 9¼. 27210-9 Pa. $9.95

THE LONG ISLAND RAIL ROAD IN EARLY PHOTOGRAPHS, Ron Ziel. Over 220 rare photos, informative text document origin (1844) and development of rail service on Long Island. Vintage views of early trains, locomotives, stations, passengers, crews, much more. Captions. 8⅞ × 11¾. 26301-0 Pa. $13.95

THE BOOK OF OLD SHIPS: From Egyptian Galleys to Clipper Ships, Henry B. Culver. Superb, authoritative history of sailing vessels, with 80 magnificent line illustrations. Galley, bark, caravel, longship, whaler, many more. Detailed, informative text on each vessel by noted naval historian. Introduction. 256pp. 5⅜ × 8½. 27332-6 Pa. $6.95

TEN BOOKS ON ARCHITECTURE, Vitruvius. The most important book ever written on architecture. Early Roman aesthetics, technology, classical orders, site selection, all other aspects. Morgan translation. 331pp. 5⅜ × 8½. 20645-9 Pa. $8.95

THE HUMAN FIGURE IN MOTION, Eadweard Muybridge. More than 4,500 stopped-action photos, in action series, showing undraped men, women, children jumping, lying down, throwing, sitting, wrestling, carrying, etc. 390pp. 7⅞ × 10⅝. 20204-6 Clothbd. $24.95

TREES OF THE EASTERN AND CENTRAL UNITED STATES AND CANADA, William M. Harlow. Best one-volume guide to 140 trees. Full descriptions, woodlore, range, etc. Over 600 illustrations. Handy size. 288pp. 4½ × 6⅜.
20395-6 Pa. $4.95

SONGS OF WESTERN BIRDS, Dr. Donald J. Borror. Complete song and call repertoire of 60 western species, including flycatchers, juncoes, cactus wrens, many more—includes fully illustrated booklet. Cassette and manual 99913-0 $8.95

GROWING AND USING HERBS AND SPICES, Milo Miloradovich. Versatile handbook provides all the information needed for cultivation and use of all the herbs and spices available in North America. 4 illustrations. Index. Glossary. 236pp. 5⅜ × 8½. 25058-X Pa. $5.95

BIG BOOK OF MAZES AND LABYRINTHS, Walter Shepherd. 50 mazes and labyrinths in all—classical, solid, ripple, and more—in one great volume. Perfect inexpensive puzzler for clever youngsters. Full solutions. 112pp. 8⅛ × 11.
22951-3 Pa. $3.95

ANATOMY: A Complete Guide for Artists, Joseph Sheppard. A master of figure drawing shows artists how to render human anatomy convincingly. Over 460 illustrations. 224pp. 8⅜ × 11¼. 27279-6 Pa. $9.95

MEDIEVAL CALLIGRAPHY: Its History and Technique, Marc Drogin. Spirited history, comprehensive instruction manual covers 13 styles (ca. 4th century thru 15th). Excellent photographs; directions for duplicating medieval techniques with modern tools. 224pp. 8⅜ × 11¼. 26142-5 Pa. $11.95

DRIED FLOWERS: How to Prepare Them, Sarah Whitlock and Martha Rankin. Complete instructions on how to use silica gel, meal and borax, perlite aggregate, sand and borax, glycerine and water to create attractive permanent flower arrangements. 12 illustrations. 32pp. 5⅜ × 8½. 21802-3 Pa. $1.00

EASY-TO-MAKE BIRD FEEDERS FOR WOODWORKERS, Scott D. Campbell. Detailed, simple-to-use guide for designing, constructing, caring for and using feeders. Text, illustrations for 12 classic and contemporary designs. 96pp. 5⅜ × 8½. 25847-5 Pa. $2.95

OLD-TIME CRAFTS AND TRADES, Peter Stockham. An 1807 book created to teach children about crafts and trades open to them as future careers. It describes in detailed, nontechnical terms 24 different occupations, among them coachmaker, gardener, hairdresser, lacemaker, shoemaker, wheelwright, copper-plate printer, milliner, trunkmaker, merchant and brewer. Finely detailed engravings illustrate each occupation. 192pp. 4⅝ × 6. 27398-9 Pa. $4.95

THE HISTORY OF UNDERCLOTHES, C. Willett Cunnington and Phyllis Cunnington. Fascinating, well-documented survey covering six centuries of English undergarments, enhanced with over 100 illustrations: 12th-century laced-up bodice, footed long drawers (1795), 19th-century bustles, 19th-century corsets for men, Victorian "bust improvers," much more. 272pp. 5⅜ × 8¼. 27124-2 Pa. $9.95

ARTS AND CRAFTS FURNITURE: The Complete Brooks Catalog of 1912, Brooks Manufacturing Co. Photos and detailed descriptions of more than 150 now very collectible furniture designs from the Arts and Crafts movement depict davenports, settees, buffets, desks, tables, chairs, bedsteads, dressers and more, all built of solid, quarter-sawed oak. Invaluable for students and enthusiasts of antiques, Americana and the decorative arts. 80pp. 6½ × 9¼. 27471-3 Pa. $7.95

HOW WE INVENTED THE AIRPLANE: An Illustrated History, Orville Wright. Fascinating firsthand account covers early experiments, construction of planes and motors, first flights, much more. Introduction and commentary by Fred C. Kelly. 76 photographs. 96pp. 8¼ × 11. 25662-6 Pa. $7.95

THE ARTS OF THE SAILOR: Knotting, Splicing and Ropework, Hervey Garrett Smith. Indispensable shipboard reference covers tools, basic knots and useful hitches; handsewing and canvas work, more. Over 100 illustrations. Delightful reading for sea lovers. 256pp. 5⅜ × 8½. 26440-8 Pa. $6.95

FRANK LLOYD WRIGHT'S FALLINGWATER: The House and Its History, Second, Revised Edition, Donald Hoffmann. A total revision—both in text and illustrations—of the standard document on Fallingwater, the boldest, most personal architectural statement of Wright's mature years, updated with valuable new material from the recently opened Frank Lloyd Wright Archives. "Fascinating"—*The New York Times*. 116 illustrations. 128pp. 9¼ × 10¾. 27430-6 Pa. $10.95

PHOTOGRAPHIC SKETCHBOOK OF THE CIVIL WAR, Alexander Gardner. 100 photos taken on field during the Civil War. Famous shots of Manassas, Harper's Ferry, Lincoln, Richmond, slave pens, etc. 244pp. 10⅝ × 8¼.
22731-6 Pa. $9.95

FIVE ACRES AND INDEPENDENCE, Maurice G. Kains. Great back-to-the-land classic explains basics of self-sufficient farming. The one book to get. 95 illustrations. 397pp. 5⅜ × 8½.
20974-1 Pa. $6.95

SONGS OF EASTERN BIRDS, Dr. Donald J. Borror. Songs and calls of 60 species most common to eastern U.S.: warblers, woodpeckers, flycatchers, thrushes, larks, many more in high-quality recording.
Cassette and manual 99912-2 $8.95

A MODERN HERBAL, Margaret Grieve. Much the fullest, most exact, most useful compilation of herbal material. Gigantic alphabetical encyclopedia, from aconite to zedoary, gives botanical information, medical properties, folklore, economic uses, much else. Indispensable to serious reader. 161 illustrations. 888pp. 6½ × 9¼. 2-vol. set. (USO)
Vol. I: 22798-7 Pa. $9.95
Vol. II: 22799-5 Pa. $9.95

HIDDEN TREASURE MAZE BOOK, Dave Phillips. Solve 34 challenging mazes accompanied by heroic tales of adventure. Evil dragons, people-eating plants, bloodthirsty giants, many more dangerous adversaries lurk at every twist and turn. 34 mazes, stories, solutions. 48pp. 8¼ × 11.
24566-7 Pa. $2.95

LETTERS OF W. A. MOZART, Wolfgang A. Mozart. Remarkable letters show bawdy wit, humor, imagination, musical insights, contemporary musical world; includes some letters from Leopold Mozart. 276pp. 5⅜ × 8½.
22859-2 Pa. $6.95

BASIC PRINCIPLES OF CLASSICAL BALLET, Agrippina Vaganova. Great Russian theoretician, teacher explains methods for teaching classical ballet. 118 illustrations. 175pp. 5⅜ × 8½.
22036-2 Pa. $3.95

THE JUMPING FROG, Mark Twain. Revenge edition. The original story of The Celebrated Jumping Frog of Calaveras County, a hapless French translation, and Twain's hilarious "retranslation" from the French. 12 illustrations. 66pp. 5⅜ × 8½.
22686-7 Pa. $3.50

BEST REMEMBERED POEMS, Martin Gardner (ed.). The 126 poems in this superb collection of 19th- and 20th-century British and American verse range from Shelley's "To a Skylark" to the impassioned "Renascence" of Edna St. Vincent Millay and to Edward Lear's whimsical "The Owl and the Pussycat." 224pp. 5⅜ × 8½.
27165-X Pa. $3.95

COMPLETE SONNETS, William Shakespeare. Over 150 exquisite poems deal with love, friendship, the tyranny of time, beauty's evanescence, death and other themes in language of remarkable power, precision and beauty. Glossary of archaic terms. 80pp. 5³⁄₁₆ × 8¼.
26686-9 Pa. $1.00

BODIES IN A BOOKSHOP, R. T. Campbell. Challenging mystery of blackmail and murder with ingenious plot and superbly drawn characters. In the best tradition of British suspense fiction. 192pp. 5⅜ × 8½.
24720-1 Pa. $5.95

THE WIT AND HUMOR OF OSCAR WILDE, Alvin Redman (ed.). More than 1,000 ripostes, paradoxes, wisecracks: Work is the curse of the drinking classes; I can resist everything except temptation; etc. 258pp. 5⅜ × 8½.   20602-5 Pa. $4.95

SHAKESPEARE LEXICON AND QUOTATION DICTIONARY, Alexander Schmidt. Full definitions, locations, shades of meaning in every word in plays and poems. More than 50,000 exact quotations. 1,485pp. 6½ × 9¼. 2-vol. set.
Vol. 1: 22726-X Pa. $15.95
Vol. 2: 22727-8 Pa. $15.95

SELECTED POEMS, Emily Dickinson. Over 100 best-known, best-loved poems by one of America's foremost poets, reprinted from authoritative early editions. No comparable edition at this price. Index of first lines. 64pp. 5³/₁₆ × 8¼.
26466-1 Pa. $1.00

CELEBRATED CASES OF JUDGE DEE (DEE GOONG AN), translated by Robert van Gulik. Authentic 18th-century Chinese detective novel; Dee and associates solve three interlocked cases. Led to van Gulik's own stories with same characters. Extensive introduction. 9 illustrations. 237pp. 5⅜ × 8½.
23337-5 Pa. $5.95

THE MALLEUS MALEFICARUM OF KRAMER AND SPRENGER, translated by Montague Summers. Full text of most important witchhunter's "bible," used by both Catholics and Protestants. 278pp. 6⅝ × 10.   22802-9 Pa. $10.95

SPANISH STORIES/CUENTOS ESPAÑOLES: A Dual-Language Book, Angel Flores (ed.). Unique format offers 13 great stories in Spanish by Cervantes, Borges, others. Faithful English translations on facing pages. 352pp. 5⅜ × 8½.
25399-6 Pa. $7.95

THE CHICAGO WORLD'S FAIR OF 1893: A Photographic Record, Stanley Appelbaum (ed.). 128 rare photos show 200 buildings, Beaux-Arts architecture, Midway, original Ferris Wheel, Edison's kinetoscope, more. Architectural emphasis; full text. 116pp. 8¼ × 11.   23990-X Pa. $9.95

OLD QUEENS, N.Y., IN EARLY PHOTOGRAPHS, Vincent F. Seyfried and William Asadorian. Over 160 rare photographs of Maspeth, Jamaica, Jackson Heights, and other areas. Vintage views of DeWitt Clinton mansion, 1939 World's Fair and more. Captions. 192pp. 8⅞ × 11.   26358-4 Pa. $12.95

CAPTURED BY THE INDIANS: 15 Firsthand Accounts, 1750–1870, Frederick Drimmer. Astounding true historical accounts of grisly torture, bloody conflicts, relentless pursuits, miraculous escapes and more, by people who lived to tell the tale. 384pp. 5⅜ × 8½.   24901-8 Pa. $7.95

THE WORLD'S GREAT SPEECHES, Lewis Copeland and Lawrence W. Lamm (eds.). Vast collection of 278 speeches of Greeks to 1970. Powerful and effective models; unique look at history. 842pp. 5⅜ × 8½.   20468-5 Pa. $12.95

THE BOOK OF THE SWORD, Sir Richard F. Burton. Great Victorian scholar/adventurer's eloquent, erudite history of the "queen of weapons"—from prehistory to early Roman Empire. Evolution and development of early swords, variations (sabre, broadsword, cutlass, scimitar, etc.), much more. 336pp. 6⅛ × 9¼. 25434-8 Pa. $8.95

AUTOBIOGRAPHY: The Story of My Experiments with Truth, Mohandas K. Gandhi. Boyhood, legal studies, purification, the growth of the Satyagraha (nonviolent protest) movement. Critical, inspiring work of the man responsible for the freedom of India. 480pp. 5⅜ × 8½. (USO) 24593-4 Pa. $6.95

CELTIC MYTHS AND LEGENDS, T. W. Rolleston. Masterful retelling of Irish and Welsh stories and tales. Cuchulain, King Arthur, Deirdre, the Grail, many more. First paperback edition. 58 full-page illustrations. 512pp. 5⅜ × 8½. 26507-2 Pa. $9.95

THE PRINCIPLES OF PSYCHOLOGY, William James. Famous long course complete, unabridged. Stream of thought, time perception, memory, experimental methods; great work decades ahead of its time. 94 figures. 1,391pp. 5⅜ × 8½. 2-vol. set.
Vol. I: 20381-6 Pa. $12.95
Vol. II: 20382-4 Pa. $12.95

THE WORLD AS WILL AND REPRESENTATION, Arthur Schopenhauer. Definitive English translation of Schopenhauer's life work, correcting more than 1,000 errors, omissions in earlier translations. Translated by E. F. J. Payne. Total of 1,269pp. 5⅜ × 8½. 2-vol. set.
Vol. 1: 21761-2 Pa. $10.95
Vol. 2: 21762-0 Pa. $11.95

MAGIC AND MYSTERY IN TIBET, Madame Alexandra David-Neel. Experiences among lamas, magicians, sages, sorcerers, Bonpa wizards. A true psychic discovery. 32 illustrations. 321pp. 5⅜ × 8½. (USO) 22682-4 Pa. $7.95

THE EGYPTIAN BOOK OF THE DEAD, E. A. Wallis Budge. Complete reproduction of Ani's papyrus, finest ever found. Full hieroglyphic text, interlinear transliteration, word-for-word translation, smooth translation. 533pp. 6½ × 9¼. 21866-X Pa. $9.95

MATHEMATICS FOR THE NONMATHEMATICIAN, Morris Kline. Detailed, college-level treatment of mathematics in cultural and historical context, with numerous exercises. Recommended Reading Lists. Tables. Numerous figures. 641pp. 5⅜ × 8½. 24823-2 Pa. $11.95

THEORY OF WING SECTIONS: Including a Summary of Airfoil Data, Ira H. Abbott and A. E. von Doenhoff. Concise compilation of subsonic aerodynamic characteristics of NACA wing sections, plus description of theory. 350pp. of tables. 693pp. 5⅜ × 8½. 60586-8 Pa. $13.95

THE RIME OF THE ANCIENT MARINER, Gustave Doré, S. T. Coleridge. Doré's finest work; 34 plates capture moods, subtleties of poem. Flawless full-size reproductions printed on facing pages with authoritative text of poem. "Beautiful. Simply beautiful."—*Publisher's Weekly*. 77pp. 9¼ × 12. 22305-1 Pa. $5.95

NORTH AMERICAN INDIAN DESIGNS FOR ARTISTS AND CRAFTSPEOPLE, Eva Wilson. Over 360 authentic copyright-free designs adapted from Navajo blankets, Hopi pottery, Sioux buffalo hides, more. Geometrics, symbolic figures, plant and animal motifs, etc. 128pp. 8⅜ × 11. (EUK) 25341-4 Pa. $6.95

SCULPTURE: Principles and Practice, Louis Slobodkin. Step-by-step approach to clay, plaster, metals, stone; classical and modern. 253 drawings, photos. 255pp. 8⅛ × 11. 22960-2 Pa. $9.95

THE INFLUENCE OF SEA POWER UPON HISTORY, 1660–1783, A. T. Mahan. Influential classic of naval history and tactics still used as text in war colleges. First paperback edition. 4 maps. 24 battle plans. 640pp. 5⅜ × 8½.
25509-3 Pa. $12.95

THE STORY OF THE TITANIC AS TOLD BY ITS SURVIVORS, Jack Winocour (ed.). What it was really like. Panic, despair, shocking inefficiency, and a little heroism. More thrilling than any fictional account. 26 illustrations. 320pp. 5⅜ × 8½. 20610-6 Pa. $7.95

FAIRY AND FOLK TALES OF THE IRISH PEASANTRY, William Butler Yeats (ed.). Treasury of 64 tales from the twilight world of Celtic myth and legend: "The Soul Cages," "The Kildare Pooka," "King O'Toole and his Goose," many more. Introduction and Notes by W. B. Yeats. 352pp. 5⅜ × 8½. 26941-8 Pa. $7.95

BUDDHIST MAHAYANA TEXTS, E. B. Cowell and Others (eds.). Superb, accurate translations of basic documents in Mahayana Buddhism, highly important in history of religions. The Buddha-karita of Asvaghosha, Larger Sukhavativyuha, more. 448pp. 5⅜ × 8½. , 25552-2 Pa. $9.95

ONE TWO THREE . . . INFINITY: Facts and Speculations of Science, George Gamow. Great physicist's fascinating, readable overview of contemporary science: number theory, relativity, fourth dimension, entropy, genes, atomic structure, much more. 128 illustrations. Index. 352pp. 5⅜ × 8½. 25664-2 Pa. $7.95

ENGINEERING IN HISTORY, Richard Shelton Kirby, et al. Broad, nontechnical survey of history's major technological advances: birth of Greek science, industrial revolution, electricity and applied science, 20th-century automation, much more. 181 illustrations. ". . . excellent . . ."—Isis. Bibliography. vii + 530pp. 5⅜ × 8¼.
26412-2 Pa. $13.95